BUILDING WITH EARTH

THE REAL GOODS SOLAR LIVING BOOKS

This Organic Life: Confessions of a Suburban Homesteader by Joan Gussow

The Beauty of Straw Bale Homes by Athena and Bill Steen

Serious Straw Bale: A Home Construction Guide for All Climates by Paul Lacinski and Michel Bergeron

The Natural House: A Complete Guide to Healthy, Energy-Efficient, Environmental Homes by Daniel D. Chiras

The New Independent Home: People and Houses that Harvest the Sun, Wind, and Water by Michael Potts

Wind Energy Basics and *Wind Power for Home & Business* by Paul Gipe

The Earth-Sheltered House: An Architect's Sketchbook by Malcolm Wells

Mortgage-Free! Radical Strategies for Home Ownership by Rob Roy

A Place in the Sun: The Evolution of the Real Goods Solar Living Center by John Schaeffer and the Collaborative Design/Construction Team

The Passive Solar House: Using Solar Design to Heat and Cool Your Home by James Kachadorian

Independent Builder: Designing & Building a House Your Own Way by Sam Clark

The Rammed Earth House by David Easton

The Straw Bale House by Athena Swentzell Steen, Bill Steen, and David Bainbridge with David Eisenberg

Real Goods Solar Living Sourcebook: The Complete Guide to Renewable Energy Technologies and Sustainable Living, 10th Edition, edited by Doug Pratt and John Schaeffer

REAL GOODS TRADING COMPANY in Ukiah, California, was founded in 1978 to make available new tools to help people live self-sufficiently and sustainably. Through seasonal catalogs, a periodical (*The Real Goods News*), a bi-annual *Solar Living Sourcebook*, as well as retail outlets and a Web site (www.realgoods.com), Real Goods provides a broad range of tools for independent living.

"Knowledge is our most important product" is the Real Goods motto. To further its mission, Real Goods has joined with Chelsea Green Publishing Company to co-create and co-publish the Real Goods Solar Living Book series. The titles in this series are written by pioneering individuals who have firsthand experience in using innovative technology to live lightly on the planet. Chelsea Green books are both practical and inspirational, and they enlarge our view of what is possible as we enter the new millennium.

Stephen Morris
President, Chelsea Green

John Schaeffer
President, Real Goods

PAULINA WOJCIECHOWSKA

BUILDING WITH EARTH

A Guide to Flexible-Form Earthbag Construction

Title page photograph is used courtesy of Hartworks, Inc.

Extract on page xi is from *Eco Design Journal,* vol. 3, no. 3, 1995.

Printed in the United States.
First printing, June 2001

04 03 02 01   1 2 3 4 5

Printed on acid-free, recycled paper.

Due to the variability of local conditions, materials, skills, site, and so forth, Chelsea Green Publishing Company and the author assume no responsibility for personal injury, property damage, or loss from actions inspired by information in this book. Always consult the manufacturer, applicable building codes, and the National Electric Code. When in doubt, ask for advice. Recommendations in this book are no substitute for the directives of professional contractors, equipment manufacturers, or federal, state, and local regulatory officials.

Many of the designations used by manufacturers and sellers to distinguish their products are claimed as trademarks. Where those designations appear in this book and Chelsea Green was aware of a trademark claim, the designations have been printed in initial capital letters.

Library of Congress Cataloging-in-Publication Data
Wojciechowska, Paulina, 1967-
    Building with earth: a guide to flexible-form earthbag construction/
      Paulina Wojciechowska.
      p. cm. — (A Real Goods solar living book)
    Includes bibliographical references and index.
    ISBN 1-890132-81-0 (alk. paper)
    1. Earth construction. I. Title. II. Series.

TH1421 .W597 2001
693'.91—dc21   2001028676

CHELSEA GREEN PUBLISHING COMPANY
Post Office Box 428
White River Junction, VT 05001
(800) 639-4099
www.chelseagreen.com

I dedicate this book to all my teachers in the world of natural building, and I thank each one of you for all the tremendous knowledge and help: Trevor Garnham of Kingston University, who taught me throughout most of my architectural education and whose idea it was to write this book; Nader Khalili and Iliona Outram of Cal-Earth, the pioneers of earthbag Superadobe construction; Athena and Bill Steeen of the Canelo Project; Tom Watson, the silent inventor of, among many things, Pumice-crete; and all the others whom I met on the way who taught me and took great care of me.

CONTENTS

Acknowledgments ix
Introduction xiii

I. EARTH ARCHITECTURE 3

Clay-Based Building Materials 5
Adobe 6
Cob 8
Rammed Earth 9
Wattle and Daub 10
Straw-Clay 10
Papercrete 11
Earthbags 12
When Are Earthbags Appropriate? 16
Extending or Recycling the
  Building 17
Earthbags Forever 19

2. USING BASIC STRUCTURES FROM NATURE TO BUILD WITH EARTH 21

Arches 22
Making the Arch 24
Vaults 24
Apses 26
Domes 26
Corbeling 26

3. GETTING STARTED: DESIGN, SITING, AND FOUNDATIONS 29

Design Considerations 29
  Locating the Building on the
    Land 29
  Landscaping 30
  Topography 30
  Orientation 31
  Utilities 32
  Building Shape 32
  Planning Ahead 33
Site Preparation: Setting Out 33
Foundations 34
  Earth-Filled Tires 39
  Rubble or Mortared Stone 39
  Gabions 40
  Dry-stone 40
  Pumice-crete 40

4. BUILDING WITH EARTHBAGS 43

Materials 44

Tools 46

Preparing the Fill 49

Filling Bags or Tubes 50

    Filling a Bag with More than Three
        People 51

    Filling Bags or Tubes with Only
        One to Three People 52

    Using Small Bags 52

Tamping 52

Keying 54

Structural Reinforcement and
    Buttressing 54

Openings 57

    Arched Openings 58

    Square Openings 60

Bond Beams 60

5. ROOFS 65

Brick or Adobe Roofs 66

Vaulted Roofs 67

Conventional Roofs 68

Water-Catchment Roofs 68

Thatched Roofs 68

Living Roofs 70

Low-cost Flat Roofs 71

Roof Insulation 72

6. WEATHERPROOFING AND FINISHES 75

Earthen Plasters 76

Application 77

Stabilization and Alternatives 78

    Vegetable Stabilizers 81

    Animal Products as Stabilizers 82

    Mineral Stabilizers: Lime 82

Lime Plasters 83

    Making Lime Putty (Slaking) 84

    Making Lime Plaster or Render 86

    Application 87

    Pozzolanic Additives to Lime
        Plaster 87

Stabilization for Waterproofing 88

Stabilization with Cement 89

Application of Stabilized Renders 91

Interior Finishes 92

Sealants 92

Paints 94

Clay Slip or *Alis* 95

    To Cook Wheat Flour Paste 96

    To Make *Alis* Clay Paint 96

    Application 97

Lime Paint or Whitewash 97

    Additives to Lime Wash 97

    Recipes for Water-Resistant
        Whitewash 97

    Some Old Limewash Recipes 98

Casein 98
  Recipe for Interior-Exterior
    Casein 99
  Application 100
Oil-Based Paints 100
Maintenance 100

7. OTHER INTERIOR WALLS,
FLOORS, AND FURNISHINGS:
BUILDING WITH CLAY 103

Finding and Analyzing Building
    Soils 103
  Jar Test 104
  Testing by Hand 105
The Right Mix 105
Thin Partitions and Ceiling
    Panels 109
Insulation 110
  Straw–Light Clay 111
  Hybrid Earthbag and Straw Bale 113
Interior Detailing 114
Earthen Floors 115
  Construction 118
  Sealants, Maintenance, and
    Repair 119
Electricity and Plumbing 120

8. THE EARTHBAG
ADVENTURE 123

Shirley Tassencourt's Domes,
    Arizona 123
Allegra Ahlquist's House, Arizona 127
House Built by Dominic Howes,
    Wisconsin 128
Sue Vaughan's House, Colorado 130
Carol Escott and Steve Kemble's
    House, the Bahamas 131
Kelly and Rosana Hart's House,
    Colorado 135
Kaki Hunter's and Doni Kiffmeyer's
    Honey House, Utah 140
The Lodge "Njaya," Malawi 142
The New House of the Yaquis,
    Mexico 143

Afterword 148
Bibliography 149
Resources 153
Index 158

ACKNOWLEDGMENTS

I would like to give special thanks to Nader Khalili and Iliona Outram, Bill and Athena Steen, and Tom Watson for their generosity in everything they gave throughout the many months of my travels, which enabled me to write this book.

I would like to express my gratitude to all those who helped me put this book together in many different ways, some of whom I mention below:

From Kingston University, I would like to thank the Green Audit Research Project for partly funding and enabling me to commence the writing of this book. Thanks to Peter Jacob, Bryan Gauld, and Sue Ann Lee. Also to Trevor Garnham for guiding me through the process. From the Green Audit room at Kingston University: Cigdem Civi, Helen Iball, and David Lawrence.

Friends in England whose help was invaluable: Flora Gathorne Hardy for her help, photographs, and immense support and belief in this book; also to Salim Khan, Bruce Ure, and Shahnawaz Khan for their technical support; and to Tim Crosskey, Henry Amos, Wasim Madbolly, and Max Jensen for the photographs; and to Adrian Bunting for the Malawi project.

Friends who contributed their stories and photographs in the United States: in Arizona, Bill and Athena Steen, also Allegra Ahlquist, Shirley Tassencourt, and Dominic Howes, who shared information as well as huge amounts of love and support throughout the whole process, together with Carol Escott and Steve Kemble, and a very big thank you for being there in those very difficult times. In California: Michael Huskey. In Colorado: Kelly and Rosana Hart. In Utah: Kaki Hunter and Doni Kiffmeyer. In New Mexico: Joseph Kennedy.

Also I would like to thank Ian Robertson in California for his support, Gene Leon for early editing, Frank Haendle in Germany for communications, Lydia Gould for all the survival help in Mexico, and Athena and Bill Steen for making that trip possible.

It is impossible to name all the people to whom I am very grateful, who were so wonderful throughout my travels, but I would like to quickly mention some not yet named, who looked after me and helped me to make my journeys: Simon Clark, John Jopling, Vanita and Alistair Sterling, Heidi Koenig, Catherine Wanek and Pete Fust, Satomi and Tom Landers, Kat Morrow, Karen Chan, Carole Crews, Cedar Rose, Lynne Elizabeth, Cassandra Adams, Leonard Littlefinger, Lance Charles, Ralf and Rina Swentzell, Arin Reeves, George Mohyla, Doc Clyne, Giovani Panza, Craig Cranic, Michael Smith, Elizabeth Lassuy, Reto Messmer, Kevin Beale, Stokely Webster, Monika Falk, and Christina and Markus Lehman.

Without my family, who provided me with constant help and support, the "journey" and the writing would have been more difficult. I would like to thank my parents, Marcjanna Sojka and Krzysztof Wojciechowski, and my stepfather, Witold Sojka.

My special thanks are extended to my partner, Christof Schwarz, who has endured many absences and who has provided constant support, encouragement, and help with writing this book, for which I am immensely grateful.

Finally, my thanks to the publishing team at Chelsea Green, especially Jim Schley, Hannah Silverstein, Ann Aspell, and Rachael Cohen, who have carefully edited the text, illustrations, and photographs.

... being surrounded by beauty allows one to put aside at least some of the burden of his or her defences against the world and to feel inwardly free. What relief! What therapy!

Unlike composition, harmony and so on, beauty rules. It is something the artist must struggle to achieve. And anyone who undertakes this struggle, with all the single-minded dedication it demands, is an artist. This love force shines from the finished product. It has nothing to do with fashion or style, and little to do with latent ability. It comes from the gift of love, and an environment so filled has a powerful healing effect, for love is the greatest healer, needing only understanding to complete it.

—Christopher Day, from "Human Structure and Geometry"

INTRODUCTION

I had a wish—to be able to go to any place on this Earth and build a shelter with the materials available to me from the surrounding environment. I had already learned about building with wood, stone, clay, straw, lime, and many combinations and permutations of all these. The material that would fill in the gap, that would give me the complete confidence that I would be able to build a house almost anywhere, was sand.

I spent most of the impressionable years of my life in Afghanistan and India, where I was surrounded by indigenous architecture. It was during that time, I suspect, that I developed a passion for honesty, modesty, and harmony in design. All the way through my architectural studies, I was drawn to what I call "primitive" architecture. By "primitive" I don't mean backward, but quite the opposite. To be *primus* means to be the first, to be at the beginning: *primary*. It is good for the mind to go back to the beginning, because the beginning of any established human activity is often its moment of greatest wonder. The original forms can teach us the fundamental principles of each invention, showing us the possibility that we might take a path other than the so-called modern way.

Most primitive buildings were constructed by an anonymous builder in response to such conditions as climate, orientation, and the availability of building materials. Building materials dictated the form of each structure. The builders were sensitive to their materials; they worked with them, not against them as so much of today's architecture seems to do.

A few years ago, I was working in an architectural office. I was always at the drawing board doing technical drawings, not really understanding the materials I was working with or why they were used. How were critical decisions made? Did the best choices depend on cost, aesthetics, or surroundings, or was the process driven by people who were just stuck in their ways, dictating by convention and code how buildings should be designed? In time, I realized that to find another way I had to learn by using my hands. It was necessary for me to experience actual construction, a time of very basic building. . . . Always start with the basics—the first principle of natural building.

An opportunity came up to participate

Facing page: Earthen structures at Cal-Earth, California.

xiii

in a workshop for people who wanted to learn to build. It was to be three intense weeks of building a house for the Othona Community Retreat in Dorset, England. The course was run by Simon Clark of Constructive Individuals, an organization in London that offers training to people who wish to build their own homes or extend or alter an existing dwelling, or who simply want to have the building process demystified. How wonderful this opportunity sounded, to build a whole house in just three weeks! I jumped at the chance.

The course covered construction techniques and also raised many ecological issues. This particular home was designed to have minimal impact on the environment by using recycled-newspaper insulation, a composting toilet system, solar electric modules, a graywater system, and a leach-field to nourish fruit trees. There it all was, and it had a name: Ecological, Sustainable, and Environmental. These were the words used for the simpler ways of building I had dreamed of. I knew that after this workshop I would know what to ask for. A whole new world was opened to me.

After the Dorset course, I became much more confident in handling tools, and many of the mysteries behind building a house were gone. I had finally experienced what it felt like to build. I got to "feel" the materials, feel what concrete is like, feel what working with wood is like, and learn how all the pieces go together. I also experienced the harsh realities of building,

which is exactly what I needed after all those years at the drawing board. It was great! These experiences reinforced my need to go away to places where people were building using basic principles and materials, places that would give me more opportunity to discover indigenous building materials and techniques, freed from the rigid commercialism of London architecture.

In the autumn of 1996, I finally embarked on my long-awaited journey. My first three-month stop was at the California Institute of Earth Art and Architecture (Cal-Earth). Set up by Iranian-born architect and author Nader Khalili and his British associate Iliona Outram, this school takes people on apprenticeship retreats for a week or more. During this time, Cal-Earth and its associates contributed to teaching me of the "magic" simplicity of earth architecture. Working with the earth empowered me to carry on, as it has empowered many other individuals. I learned the basics of earthbag construction (called Superadobe at Cal-Earth) and spent time researching this method, trying to push its limitations and explore its possibilities.

To simplify is the aim. What a joy to learn and to fulfill my dreams! I spent a wonderful three months, studying as well as building my own retreat. For the first time in my life I could design and build a house with no restrictions from anyone. I was free and I felt free and therefore I expressed freedom.

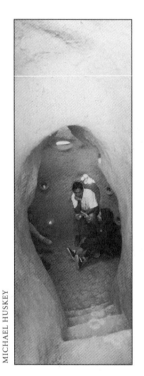

MICHAEL HUSKEY

Interior of the author's Earthmother Dwelling retreat.

Here is an extract from my diary, describing my thoughts when I first arrived at Cal-Earth:

It's early morning as I step outside the house. Mist covers the bottom of the mountains, silhouettes of shabby Joshua trees, a tree that stands so still all the time. The sand is lit up, the mountain towers above the mist. The cold is sharp, the sun is bright. In the foreground I see domes and desert architecture, like sand dunes. The landscape looks and feels right. Instead of destroying the view of the mountain beyond and Joshua trees around and the vast openness, the structures enhance the view. I do not feel disturbed by their presence. Quite the opposite; I am happy they exist and have become an integral part of this landscape. I walk toward them, at first looking around the outside. I realize that they are sunken into the ground. They feel very solid and permanent. Yes, they have this permanence and belonging about them, a permanence that a timber house does not have. Once I walk inside, they feel cool. They have soaked up the night's cold air. I am drawn toward the one that looks like an animal. It has a fireplace and a small niche to hide in. Ideas flow into my mind of a house I would like to build right now for myself, with a network of rooms all interconnecting, partly underground. The outside is very intriguing as well. I imagine patterns forming from the way the plaster has been applied and outlined. After a while, I proceed to the other attractive-looking earth shelter. This one is totally covered in earth, or so it appears. It has tiny circular windows going all around at the lower level, great for children if they want to look out. The light at sunrise is amazing; it floods in though one larger window. Because of the partially white wall, the room looks bright. The primitive paintings on the wall stand out. The light covers the horizon; the sun is a little higher, but the air is still sharp. I sit in the sun's rays to warm myself.

After some time I get up and walk into a brick dome. The light beautifully percolates through the bricks. The top of the dome is open. This one feels the coldest so far, probably

The author's earthbag retreat before plastering.

The author's retreat during plastering.

because none of the openings are glazed, and I feel the air moving through. I walk to the center. The acoustics are amazing. The sounds that I make with my feet bounce around and seem very loud. What a great place to play music this would be. After a short pause, I proceed to the Three Vault House. Whoa! Rectangular rooms, white walls—it feels like a chapel. A lot of different-sized openings and niches in the walls. A cooling tower under construction. A very large space in comparison to the others. The light reflects off the walls in a soothing way. It feels airy and open. There are more buildings here than I expected— but what did I expect?

My desire to build using the earth, utilizing the earthbag technique, was very strong. There were many reasons for this, including a longing to research an unknown material, to understand and celebrate its possibilities. I also wanted to find out what it takes for a woman with few manual skills and little strength to build single-handedly a shelter for herself and her children. I wanted to indulge in the freedom of using earthen materials, having learned about the structural form of the arch, which allows the use of only earthbags for the whole structure, with no reliance upon wood.

My passion for the arch as a structural form has little to do with the ways symmetrical domes or vaults have been used traditionally, for example in Islamic architecture. Rather, this passion has to do with the prevalence of arched forms in nature. I did not necessarily want to imitate nature. I wanted to experience the freedom that nature appears to possess. I had experienced so many constraints in the world I came from that I wanted to escape rules and regularity. I wanted to become *free*.

So my ideas just came as I went along. Later, looking back, I could translate these ideas into theories about design and construction, but that is not how it started. The accompanying drawings and photos give a sense of how I went about designing and building the retreat.

To begin with, I spent many days looking around the site thinking about where to place my tiny retreat. I found a lovely spot, away from other structures. A lonely Joshua tree stood in a clearing of sorts. I wanted to be close to the tree, having it in front of my courtyard, with the entrance facing east, because the strong winds came from the west. I wanted the retreat to provide a place to sit facing the courtyard and the Joshua tree.

When planning the retreat, I tried to figure out what I would like inside it, never restricting my imagination. I knew that people who stayed here would need a place for their luggage, a place to sleep, and perhaps a place for a child to sleep. So I designed for a couple and their child. A fireplace would be used to keep warm and to cook on. Then I thought of the views, the light, sunrise and sunset, and the lookout points for observing who was approaching. The seating areas were to be niches so three or four people could sit facing each other. And a child who lay sleeping in a niche would be able to see the fire burning opposite, for comfort. I wanted the ceiling to be high, since we often judge a space by its volume. Although the main room was only 10 feet (3 meters) in diameter, ten people could sit comfortably inside without feeling claustrophobic.

An earthbag dome sunken into the ground creates and encloses a space. Basing your design upon that central enclosure, you can add or take away as you please and as the structure allows. Building with earthbags is not like digging a cave, scraping away at the earth to hollow out the shape of the building, and it's not like a wooden or concrete house where everything has to be precisely placed to accommodate manufactured materials. Building with earth, you can add as well as remove material to create the shapes you desire.

The earthbag allowed me a great deal of freedom. I wanted to celebrate this. As I worked with this totally fluid material, earth or sand in bags, I wanted the materials to lead me. I did not want to make it imitate another kind of building. I wanted to set the dome free, to listen to it. I was building with love, creating something that felt right. I believe that all buildings should be designed and built with sensitivity. To me, love has become the most important factor in designing. In fact, I now see that designing and building are like sculpting. When sculptors carve rock, they listen to the rock telling them what it needs to become. Designing and building a dwelling is likewise about understanding the material, the needs and passions of the inhabitants, and the climate and other characteristics of the site. The challenge is to be in harmony with your environment, and most of all to feel passion during the process.

To me this is what contributed to my retreat's soul.

From the moment I started to build, I recognized that it was essential to maintain trust in the material. Without this total trust, you fall back upon narrow ideas. To push any process forward, you need trust. Unless you build with feeling, you will not feel true contentment with the finished product. When you merge the natural and the human-made environment into one, when you listen to the sun and the wind and the natural forces all around,

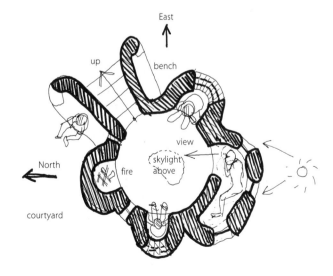

MICHAEL HUSKEY

The author (in foreground) with Iliona Outram (behind) and other Cal-Earth visitors inside the Earthmother Dwelling.

the building can be an organic extension of the land and the outcome of a marriage between wind, sun, and the soul of the one who dwells there.

At the moment, my ideal house is one that lives in such harmony with its environment. It is a house that is difficult to notice, like an animal that blends into its surroundings. So many houses appear like warts in our landscape. When you drive through the countryside, how much nicer it would be if you couldn't see the houses, if they blended in harmoniously, like the houses that climb the hillsides in Afghanistan, made of the same earth as those hillsides. Only at night, when the lights come on, do you see the extent of the developments.

In building my "Earthmother Dwelling Retreat" at Cal-Earth, these are some of the sensitivities that I brought to the process.

In this book, I will give a thorough introduction to earthbag construction and a basic introduction to some other forms of "alternative architecture." You have to understand that this is a drop in the ocean. I can only introduce the concepts, sharing some of the lessons I have learned and sights that I have seen. Numerous books have been written on many of the natural construction techniques, some of which can be ordered from the organizations and bookstores listed in the resource section. However, to my knowledge, no one has written a book-length work about the earthbag construction technique, only short articles for various magazines. This was my primary reason for putting pen to paper—to provide some of the theoretical knowledge that I gathered in the places to which my research took me. That theoretical foundation is necessary to begin construction, but the reader must remember that no amount of theory can teach as much as your own hands. Happy experimenting!

BUILDING WITH EARTH

Earth Architecture

Since the earliest times, people have lived in the earth, taking up residence in existing structures or forming and sculpting earth around them according to their needs. In terms of growth and development, indigenous communities usually lived within the limits of their ecosystems. Nature, technology, and culture maintained a balance.

Until the industrial revolution, most of the world's people housed themselves in earthen architecture (Khalili 1986, 58). Even today, it is estimated that a third of the human population lives in houses constructed of unbaked earth. But, during the industrial age, the use of engines and fossil fuels expanded the limits of local ecosystems. Resources from distant regions were brought together in the process of mass production. Industry on an unimaginable scale transformed the landscape, while the goods manufactured on assembly lines transformed our values. Yet for a long time the environmental consequences of modern design seemed remote. At the beginning of the twentieth century, architects were inspired by machinery, not by nature. Modernists saw buildings as isolated objects, not as part of larger systems or communities. The designs of the industrialized world have developed to depend on materials and technologies beyond the limits of what local ecosystems provide. It may seem as if technology has given people the freedom to override the laws of nature. But if we use that freedom, we must take responsibility for making these choices. Today, our global technologies are depleting the Earth's resources, darkening the water and skies with waste, and endangering the diversity of life. Can we find a way of life that will re-create a balance between nature, humanity, and technology?

Above: An afternoon snooze in the city of Petra, Jordan.

Facing page: The ancient city of Petra, Jordan, carved out of sandstone.

Today, timber, steel, and cement are not readily available in many parts of the world. They must be transported from thousands of miles away. For the people who live in those regions, it would make more sense to build their own houses out of what is "beneath their feet" and to use the materials within their reach. Yet in the less industrialized countries, where modern building materials are used by the privileged few, poor people often look down upon the ancient construction methods and scorn the earth as a building material. In response to this, some designers and builders, including the late Egyptian architect Hassan Fathy, have attempted to revive ancient techniques, building earth houses for the poor as well as for the rich.

Earth has been used to build on mountains, cliffs, marshlands, and the harshest of deserts. With a suitable mix of ingredients and appropriate design, earth can be used to build almost anywhere in the world. Why not continue to use earth and other natural materials where it is appropriate to do so?

One of the reasons for the revival of earth architecture and for the sudden rise of interest in alternative ways of building is the change in people's level of consciousness. A growing movement to promote "natural," "environmental," "sustainable" understanding is now trying to make people aware of the devastation that we have caused on our planet. Among those designers and builders involved in build-ing houses, the use of local materials is becoming increasingly important, among many other environmental aspects. Using local materials not only saves energy and resources, it gives builders and dwellers a sense that they are grounded, a sense of belonging, which is missing when "foreign" materials are used. In the southwestern United States, for example, the Pueblo people have traditionally used adobe, a sun-dried clay brick, because sun and clay are easily acquired and worked with. In the northeastern United States, houses were traditionally made of wood, not because clay wasn't there, but because wood was plentiful and easily available. In England, particularly in Devon, cob was the popular building material, as the soil was perfect for forming the loaf-shaped lumps that constitute the basic building block in this method. Sadly, people have almost stopped building earth houses in England, now that bricks and concrete are cheap and available.

In the United States today, builders are gradually returning to older and more natural construction techniques. People are learning from ancient European earth-building traditions such as cob and wattle and daub, as well as from the traditions of the Pueblo Indians adapted from the straw-clay adobe building technique that was brought by the Spanish a few centuries ago. New methods, such as straw bale and earthbag construction, which combine the benefits of a variety of traditional techniques with elements of modern technology, permit people to build dwellings that are appropriate for the sites and climates where they are built. It is beyond the scope of this book to provide detailed descriptions of each natural building technique, and the books listed in the bibliography provide more comprehensive information on adobe, cob, rammed earth, straw bale, and other methods. Here I will give brief introductions to several traditional techniques that are especially useful in combination with earthbag construction.

CLAY-BASED BUILDING MATERIALS

A critical ingredient in durable, resilient earth for building is clay. In response to cultural, climatic, and geographical differences throughout the world, many variations of clay-based earth architecture have been developed. These techniques can be traced back thousands of years; for example, traces of mud walls from more than two thousand years ago were found at the Tel Dor excavations in Israel (Stern 1994, 133). Parts of the Great Wall of China are built out of earth and are still standing today. In many cultures, clay has long been considered a magical material. It is written in holy books and poems that humans themselves were created from clay.

In the Pueblo cultures of the American Southwest, one of the deities is the Pueblo Clay Lady, who is said to live in each piece of clay pottery made in the traditional manner: She inspires and advises her potters what to do while working with clay.

There is a Tewa prayer, of which the Santa Clara Pueblo poet and potter Nora Naranjo-Morse says (Swan & Swan 1996), "This prayer continually renews our relationship to the earth, her gifts, and [the people]."

Clay Mother,
I have come to the center of your
*   adobe,*
feed and clothe me
and in the end you will absorb me
into your center.
However far you travel,
do not go crying.

Clay is the result of the chemical weathering of rock and silicates such as feldspar, quartz, and mica. The diameter of clay grains is smaller than two one-thousandths of a millimeter. There are several types of clay, but the most commonly found are *kaolin* and *montmorillonite*. With an electron microscope, one can see that these materials are wafer-thin, foliated, and scaly crystals.

Earth alone (without the use of forms) can be used for construction only if it contains some kind of stabilizing element, which in industrialized architecture is often cement. In traditional natural building, we use clay as the binding material, due to the cohesive properties of the clay molecules. These molecules are attracted to each other, therefore producing a strong bond. If clay particles are well distributed throughout the soil, they form a coating around the particles of silt, sand, straw, and gravel or other filler used, effectively binding them together.

On drying, the swollen clay shrinks unevenly and causes shrinkage cracks. The more water that is absorbed by the clay, the larger the cracks will be after drying. Each type of clay has a different chemical compositions, but above all they vary in their water-absorbing qualities. Kaolin absorbs water the least, while montmorillonite can absorb seven times as much water, and can swell to sixteen times its volume. Working the clay with your hands enables the clay particles to pack together in denser, parallel layers, creating stronger binding force. As a result, the tensile and compressive strength is greater, up to 20 percent more than in mechanically compressed blocks.

Many different types of traditional earth construction require some clay as the binder for cohesion, including adobe, cob, rammed earth, wattle and daub, and blends of clay with straw or other fibers. Because of serious, building-related environmental problems in industrialized countries, in recent years we have seen a dramatic revival of traditional building techniques based on clay. Clay is nontoxic, recyclable, and easily available in many parts of the world. Combined with sand, gravel, and natural fibers such as straw and wood, clay is again becoming popular as a base for interior and exterior plasters and construction materials. This is largely due to cost and energy savings, and to the way that houses built with clay-based materials are more aesthetically pleasing and healthier to live in, especially for chemically sensitive individuals.

See chapter 7 for more detailed information about working with clay as a building material.

ADOBE

Adobe blocks are sun-dried mud bricks that can be stuck together with mud mortar to create thick walls. They have been used for thousands of years in North Africa, South America, Asia, and the Middle East, and were brought into the southwestern United States by the Spanish. The Spanish learned about adobe construction from the Egyptians, who still use this ancient building technique. Using the arch, dome, and vault, it became possible to create houses using only earth (Fathy 1986). Many magnificent large adobe structures are still in use in Afric, Asia, and the Middle East.

While in most of the world's countries adobe is used for the poor, in the southwestern United States it is increasingly fashionable with the very rich to live in a healthy, natural house. Because adobe is very labor intensive, it is very expensive to pay someone else to build and finish an adobe house. But, because the materials—the earth at your feet—are practically free, it is very inexpensive to build with adobe if you do the work yourself.

The Taos Indians of New Mexico have a long tradition of earth building using adobe. Adobe houses are part of the culture of the Pueblo people. They go together like Eskimos and igloos. During the persecutions by the Spanish and by the Americans the adobe walls protected the Pueblo people and allowed them to keep many of their spiritual beliefs, attitudes, and practices to themselves.

The beauty of earth architecture is that it participates in the natural environment. It is part of a continuous cycle, unlike contemporary industrialized architecture, where structures are built to stand independent of and unaffected by their surroundings. Build, live and die, build, live ... Adobe houses, like other earth houses, have to be cared for continuously and when their useful life is over, they are given the respect of being allowed to melt back into the earth. For example, a house is boarded up after the death of its owner. New people moving in will mold new adobes from the material remaining, and the cycle will continue.

Adobe structures flow out of the earth, and it is often difficult to see where the ground stops and the buildings begin. By using adobe to build the walls of a house and cob to sculpt the interior, beautiful, curved forms can be fairly quickly constructed that provide very inviting living space while also providing the mass

Adobe houses in Taos Pueblo, New Mexico.

For adobe houses to live on we have to take care of them, like you would take care of a child. You coat the child, we plaster the adobe houses, so they can stand up. When you live in a house the house is almost like a part of you because you live in it and it lives with you. When you keep the house warm it will keep you warm. If you die the house will also die with you. When you build a house, you build the house with what you're actually standing on. Once upon a time there might have been a house there also, but the recycling of the adobe material is almost like building what had been there once upon a time.

—Joe Martinez, Taos Pueblo,
quoted in *At Home with Mother Earth*
(Feat of Clay, 1995)

Adobe house of Hassan Fathy, Egypt.

A boy making adobes in Peru.

needed for a building to perform well thermally. Ovens and seating as well as walls can be created using adobe finished with an earthen plaster. (See chapter 7 for more discussion of how to use adobe in conjunction with other earthen techniques.)

The traditional method of adobe preparation is a highly laborious process. The strength and durability of the finished dwelling depends upon the quality of the bricks. Adobe bricks are produced by putting the appropriate soil, clay, and straw mixture into a mold where the mix is worked lightly by hand then quickly removed. The mold must be clean and wet to ease removal of the formed brick. Adobes can also be made without forms, but a large quantity of mortar must be used to smooth out the unevenness of the joints.

Adobe is sometimes criticized for being a very soft material, but adobe construction is a system. That is, no single brick is subjected to intensive pressure, because the overall wall, which is stronger than its individual parts, carries the weight of the roof. To add to its strength, adobe can be reinforced with a diversity of fibers.

COB

Cob is like an adobe mix with as much straw as the mud mixture can accept before it fails to bind. That is, subsoil containing clay is mixed with straw and water and brought to a suitable consistency by kneading or treading. The lumps of earth (or "cobs") are then placed in horizontal layers to form a mass wall. The bulk of a cob structure does not always consist solely of cob mixture. Cob that contains gravel or rubble can be sculpted into walls, making the whole structure more resistant to moisture, allowing more air to circulate inside, and keeping it drier.

Cob was traditionally a popular building material in the western parts of Britain, mostly Devon and the southwestern regions, because the soils of that region are among the best in Britain for earth construction. Most soils there contain a good proportion of clays that are fairly coarse and therefore do not expand and contract extensively and which provide adequate cohesion. Secondly, these soils are usually found to contain a well-distributed range of aggregates, from coarse gravel to fine sands and silts. A good, well-

graded subsoil mixed with plenty of straw requires no other additives to make good cob for building.

The durability of a cob house depends on how much energy is put into it as well as what is in the mix. If a cob walls fails it is usually not the fault of the material but of the builder. Like adobe houses, if the cob house is loved and cared for it will last for a long time.

RAMMED EARTH

This is another age-old technique that utilizes only the earth to create thick, durable walls, which can be load-bearing, low-cost, heat-storing, and recyclable. Rammed earth structures can be built in a variety of climates and will last for hundreds of years. The construction procedure is simple. A mixture of earth is rammed between wooden forms. The forms are removed, creating thick walls that need no external finishes. The most basic type of rammed

earth structure can be made if a minimum of 5 percent clay is present to bind the soil and if wood or other material is available to make temporary forms.

Rammed earth is another ancient earth-building technique currently being revived in many parts of the world. The Great Wall of China is partially built out of rammed earth, and this technique has been used in Yemen to build structures as high as seven stories. In eighteenth-century France, a pioneering architect named Francois Cointeraux tried to revive rammed earth construction with little success due to fear of competition among other builders. Currently in France, the use of earth as a building material is being revived by the organization CRATerre. In

Cob fireplace sculpted by Kiko Denzer at the Black Range Lodge, Kingston, New Mexico.

Rammed earth house in Arizona.

Australia, it is a popular alternative building technique, and its modern application in the United States has been pioneered by David Easton, who updated Cointeraux's techniques with improved engineering, sophisticated forms, and innovative design to make rammed earth cost-competitive with conventional construction. Modern equipment speeds the process. The soil is mixed on-site and then poured into the wooden forms set up on top of an appropriate foundation (usually stone or concrete). An earth mixture with a moisture content of 10 percent is then rammed in 6- to 8-inch layers using pneumatic or hand tampers. The forms are removed, revealing a 2-foot-thick wall that is then complete. A concrete bond beam is poured for the top of the wall on which the roof will sit. Even at its simplest, rammed earth requires more complex technology than adobe or cob but can be used to raise massive walls in a shorter period of time (Easton 1996).

WATTLE AND DAUB

In Britain, wattle and daub was widely used in the construction of internal walls and ceilings and also for external walls of houses. Wattle-and-daub panels in timber-framed houses were in common use until the eighteenth century.

The wattle (branches) act as support for mud plaster (daub). Oak or other timber stakes are installed vertically into a frame woven out of willows or other flexible wood and covered with a heavy mixture of

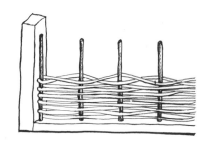

A wattle panel under construction.

clay and straw. When dry, the surface can be plastered with a mixture of lime, sand, and animal hair and painted with whitewash. Chapter 6 describes lime plasters and finishes for use on any earthen wall.

STRAW-CLAY

Straw-clay is the general term used for any building material that is made out of straw and clay (with some sand to reduce cracking and increase mass) but does not fit into the traditional adobe or cob category. Although the mixture can be very similar to adobe or cob, the main difference is the greater quantities of fiber. If the binding clay is diluted with more water to a creamy liquid consistency prior to mixing with fibers, it becomes "light clay." The most popular uses of straw-clay mixes include straw-clay blocks (straw coated with light clay rammed into formwork), thin interior or exterior walls reinforced with bamboo or branches, ceiling infill between beams or floors, or straw-clay panels for thermal or acoustic insulation. Higher density of straw allows for better insulation for roofs, floors, or the insulating layer on an

Advantages of Adding Fiber

Some of the many advantages of adding fibers such as straw to an earthen mix are:

- controlling the shrinkage cracks
- increasing tensile strength
- improving insulation value

There are also many advantages of coating natural fibers with earth:

- increasing compressive strength
- providing fire resistance
- improving water resistance
- improving insect resistance

BILL STEEN

earthbag wall. Other combinations of fiber, clay, and sand are probably being developed, as the possibilities are endless. See chapter 7 for more detailed discussion of straw-clay mixtures.

Earthen materials and natural fibers work together extremely well, preserving and protecting one another, and combining the earth's thermal mass with the fiber's insulative value. Bill and Athena Steen, who have been using straw-clay techniques in Mexico, have said "they can be built from predominantly local materials in whatever combination best matches the local climate. Like the rest of life, building can be much more fulfilling when founded on a good relationship. For us, … combining earth with natural fibers has led to an unfolding of options and possibilities that would not be open to us if we were to remain simply straw bale builders."

The main disadvantage of adding fiber to an earthen mix is the reduction in the material's thermal properties. The more straw, the less thermal mass the structure will have. This is only a disadvantage if the surface is needed for passive solar heat gain or cooling, in which case less straw and more sand can be added to the mixture.

PAPERCRETE

"Papercrete," or "fibrous cement," is an experimental technique recently reinvented independently and pioneered by Mike McCain in Alamosa, Colorado, and Eric Patterson of Silver City, New Mexico. Papercrete is a type of industrial-strength

Straw-clay block construction of the office headquarters for the Save the Children Foundation in Cuidad Obregon, Mexico. Straw-clay can be used to make large blocks to construct whole walls, wall or ceiling infill between framing or floors, large ceiling panels made by inserting bamboo or branches as reinforcement, or even rolls and arches.

papier-mâché used to make large blocks to construct houses as well as a plaster for covering them. It has been successfully used in earthbag projects as a thick plaster. If applied in several 2-inch (50 millimeter) layers, papercrete will contribute to the insulation value of an earthbag house. According to Gordon Solberg in an *Earth Quarterly* special issue on paper houses (see the bibliography), papercrete's R-value is 2.8 per inch. When dry, papercrete is lightweight, holds its shape well, and is remarkably strong. Papercrete is also durable when wet, although not waterproof, and in very damp climates will need to be sealed with a waterproof layer. For instance, a coat of tar has been used successfully as waterproofing on top of a papercrete structure in Colorado.

To make papercrete take a large mixing container and soak old newspapers and magazines until they are soft. Mix in cement or lime and sand for greater strength. The recipe for the papercrete plaster mix Kelly and Rosana used on their house (see chapter 8) is as follows:

First coat (more insulative): 1 part paper to 1 part cement + water
Second coat (more waterproof): 1 part cement to 1 part lime to 8 parts sand + water

To make papercrete blocks for construction, Gordon Solberg recommends mixing together a "soup" of 60 percent paper, 30 percent screened earth or sand, and 10 percent cement. This mixture is then poured into forms to make blocks and can also be used as a plaster or mortar.

In *Adobe Journal* (nos. 12 & 13) Mike McCain described a procedure for making large quantities of papercrete. Fill a tank with water and add magazines and newspaper. Start mixing, and when the mixture turns into a slurry, add 8 to 9½ one-gallon buckets ⅔ full (or 6 shovels-full, approximately) of sandy soil that has been sieved using chickenwire to eliminate larger rocks. Then add one 94-pound bag of cement to the mix. The amount of water needed can be estimated by watching as the cement is absorbed by the paper fibers. Be sure that the water is evenly distributed throughout the mix, as any excess water will separate. By weight, 20 percent of the mixture should be one-half sand, one-quarter paper, and one-quarter cement, and the remaining 80 percent of the mixture should be water.

Mike has devised a faster and easier way of mixing by mounting a barrel with an overturned lawnmower blade in the bottom above the axle of a cart. He uses a drive shaft to turn the lawnmower blade when the wheels of the cart turn. The mixer can be towed behind a car or truck, or even a horse.

Earthbags

Earthbags are textile bags or tubes filled with earth (sometimes sand or gravel), rammed to a very solid mass, then used to construct foundations, walls, and domes.

This method of construction is rising in popularity among natural and alternative builders, especially in the United States. This technique is essentially a flexible-form variation of rammed earth. The bags are permanent forms that allow you to ram or tamp the earth to create thick earthen walls, symmetrical arches, vaults, or domes, and freeform landscape features; and to sculpt forms the way a potter molds coils of clay, blending your structure into the landscape.

The earthbag technique requires few skills and tools other than a shovel, and can be used on almost any land, in any location. When built properly, earthbag walls are extremely strong. They are most advantageous in remote areas with no wood for a frame and no clay for a cob or adobe building, since the use of bags as a container allows the builder to utilize a wide range of soils, from unstabilized earth or sand directly from the site to soils with a high clay content, or even gravel. As a result, costly materials such as cement and steel can largely be avoided. Earthbags can even be used in areas prone to flooding and periodic wet conditions.

If an arch is used as the primary structural element—for example, a dome for the central structure with self-supporting arches for openings—no wood is necessary in the construction, thereby avoiding the deforestation so widespread in the United States and around the globe.

Earthbag buildings are low in materials cost, but intensive in labor, albeit less labor-intensive than adobe, cob, or rammed earth. Earthbags can be used in areas with limited technology and low income, but where people are willing to work on constructing buildings for themselves. The bags are cheap and easily transported, so this technique can be used for disaster-relief housing.

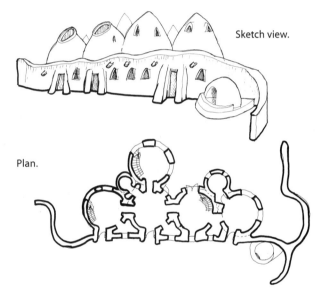

Sketch view.

Plan.

Interconnected spaces can be arranged in response to the site and solar exposure.

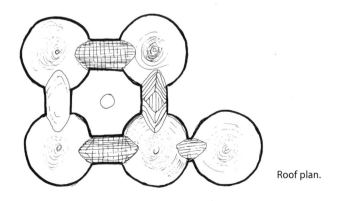

Roof plan.

An earthbag "dome" is rounded overall, but not necessarily symmetrical. The height of the dome proportionally exceeds its diameter. The difference between earthbag domes and the high-tech geodesic domes invented by Buckminster Fuller is that earthbag domes can never have large spans. An earthbag house can be anything from a single dome to a whole village of domes interconnected by vaults. Arched openings can form entrances into other spaces, niches, and apses. If a large earthbag house is desired, many domed structures can be built, each one joined to the next using a small vault. New openings can be easily cut out and extra rooms added as required. Such structures grow organically, each addition buttressing the next one. Chapter 2 explains the principles of arches, domes, and vaults that you will need to know to build these structures with earthbags.

Through recent history, sandbags and earthbags have been used for varied purposes, mostly in emergency relief work, for example to provide erosion or flood control, when filled with earth or pumped full of concrete or soil cement, then used for fast construction of embankments. Earthbags have also been used by archaeologists, to aid in structural support of collapsing walls; by armies, to create bunkers and air-

Structural and Seismic Testing

Between 1993 and 1995, three of the experimental earthbag structures at Cal-Earth passed structural tests approved by the International Conference of Building Officials (ICBO), leading to a building permit for the Hesperia Museum and Nature Center in California in March 1996, the first to be granted under the California Building Code for earthbag construction. A school initially designed by Nader Khalili and Iliona Outram in Nevada is currently under construction with code approval (Outram 1996).

The tests carried out for the ICBO included a live-load test and dynamic and static-loading tests. For some of the tests, cables were wrapped around three buildings of different designs, and hydraulic jacks were used to pull 3,000 pounds of cumulative pressure every 15 minutes. Nader Khalili explains, "On our superadobe prototype we went from 3,000 to 26,000 pounds of continuous stress and held it there for hours. This was beyond any required code for any building within fifty miles. The inspectors found no cracks or movement, so now this method is approved for all types of buildings, from residential to commercial." (*Designer/builder* June 1996.)

In high wind and earthquake areas, symmetrical buildings tend to be less heavily damaged. The form of a dome is more likely to absorb an earthquake's jolts and spread the shock equally through the structure, and the weight and curve of a dome deflects winds, allowing them to

raid shelters; and by landscapers, to create free-form retaining or enclosure walls.

According to J. F. Kennedy (1999, 82), German architect Frei Otto experimented in the 1960s with using earthbags for building. In 1978 a team from the Forschungslabor für Experimentelles Bauen (FEB), the Research Laboratory for Experimental Building at the University of Kassel in Germany, led by professor, architect, and author Gernot Minke (*Lehmbau–Handbuch*, 1997), set up an experimental earthbag project. This experiment was followed by a joint research project sponsored by the FEB, the Center for Appropriate Technology (CEMAT), and the University Francisco Marroquin in Guatemala, which attempted to develop an earthquake-proof system. Long cotton tubes, dipped in lime wash to prevent the bags from deterioration, were filled with volcanic earth (pumice) and stacked between bamboo poles, serving as a prototype for a single-story house. The poles were tied with wire every fifth course and fastened to the foundation below and a bond beam above, creating movable but earthquake-resistant walls. Further research was carried out by Minke at the university through extensive experiments with earth-filled sacks and tubes to create various structures, including domes.

---

wrap around the building rather than lifting it (Muller 1993). By contrast, vaults have a very poor earthquake resistance. This is due to the torque oscillations in structures, caused by the strong increase in seismic accelerations transferred from the terrain to the foundations when at different levels the centers of gravity do not coincide with the centers of torsion (Houben and Guillaud, 313). For seismic areas, Nader Khalili has developed a foundation system where the base of the dome is isolated from the slab it sits on by a layer of sand, therefore enabling the structure to move freely during an earthquake, (see chapter 3 for more on foundations).

Above: In 1993 the sandbag and barbed wire system was analyzed at Cal-Polytech, San Luis Obispo, where testing included observations on a scale model on a seismic table (Outram 1996, 58). The earthbag structures tested were constructed out of unstabilized earth, with barbed wire between each course, an adobe plaster finish inside and outside, and metal strapping loosely netted about the structure to contain bursting forces. The intersections were riveted or bolted together, and four zones of diagonal "X" strapping were added for resistance to shear forces (Kennedy 1994, 19).

An experimental dome at Cal-Earth, has stood for five years without interior plaster. Though the polypropylene bags have degraded from ultraviolet light, the compacted earth, which has a low clay content, is falling apart as could be expected.

The recent trend in using earthbag technology in building homes is largely due to the pioneering work carried out by the Californian Institute of Earth Art and Architecture (Cal-Earth) in the Mojave Desert, which was set up by Nader Khalili, Iranian-born architect and author, who calls his building technique "Superadobe." Since 1990, the team at Cal-Earth, in collaboration with the city of Hesperia and many researchers and associates, has been investigating earthbag construction and developing its applications, from straight walls to domed structures. To satisfy a desire (or even an obsession) to avoid wood, they have managed to create stable dome-shaped structures using the corbeling method (see chapter 2). This means that with no materials other than bags, barbed wire, and local earth, you can build yourself a shelter anywhere in the world.

Many people who have attended workshops at Cal-Earth have started to spread their knowledge. The first fully functioning and lived-in earthbag dome was built in Arizona for Shirley Tassencourt by Dominic Howes, who went on to build other round, domed, and straight-walled or square earthbag houses and water-storage tanks. Many others have followed, including a domed structure built in Utah by Kaki Hunter and Doni Kiffmeyer of OK OK OK Productions; a three-vault house built by Cal-Earth apprentices in Sarmiento, Mexico; a development built by Mara Cranic in Baja California; a two-story house, the ground floor constructed of earthbags and the first floor of timber, built in the Bahamas by Carol Escott and Steve Kemble of Sustainable Systems Support; a hybrid earthbag and pumice-filled-bag house built by Kelly and Rosana Hart of HartWorks; and Joseph Kennedy's experimental work in South Africa with bags containing high proportions of clay, cement bond beams, and bag additions to existing structures.

Earthbags are also becoming popular among natural builders as foundations, filled with gravel, sand, and/or earth, beneath straw bale and cob buildings. Earthbag structures can range from emergency refugee housing for the poor to elaborate, modern residences complete with plumbing and electricity.

WHEN ARE EARTHBAGS APPROPRIATE?

The bags are used only as a temporary formwork for ramming the earth, before the plaster is applied. It is actually the plaster that should be considered a permanent enclosure or casing. The materials used to fill the bags can range from very loose gravel, pumice, or sand to a more compactable soil, which might contain varying amounts of clay. The weaker the fill material (meaning it contains little binding strength such as clay) the stronger the bag material must be. If the soil has a high clay content, bags might not be necessary to contain the earthen mixture (see the discussion of soil testing in chapter 7). In that case, other construction techniques such as adobe or cob might be more appropriate. When earthbags are used, especially in flood areas, care must be taken that the lower courses of the wall do not contain clay, which wicks moisture. If the higher courses contain clay, they need to be tamped or rammed well to reduce the likelihood that insufficiently compacted clay will absorb moisture.

Extending or Recycling the Building

It is helpful to plan for future extensions of an earthbag building during the initial design stage in order to prevent awkward redesign problems later and to avoid the need to rethink all the openings and utilities. For example, anticipating a future opening by building an arch in that location and filling it in with nonstructural earthbags or another material such as straw bales or straw-clay mixture can save much time when you decide to cut through the original wall to make an addition. But if the earth in the bags is compactable and not just pure sand, which will simply spill out when the wall is breached, it is also possible to make openings later if really necessary. The plaster-

work can be hacked off, exposing the bags. Cut the bags with scissors or a scalpel, and scoop out the hard, rammed earth. The bags can then be resealed using nails like tailor pins or sewn together with wire, and the plaster reapplied. This is a time-consuming process, and care must be taken that the opening cut out is in the shape of a steep arch for structural stability (see the discussion of openings in chapter 4). If a square opening is cut out, the opening will not support rows of bags above. Square openings must therefore be the height of the whole wall. Better yet, plan ahead and build in arched openings that can be accessed later by removing the finish plaster. The cheapest and most flexible option is to plan future extensions at the outset. Do not be afraid to start small and expand later.

An earthbag structure can also be completely recycled. If the bags were filled with pure earth from the land the building is standing on, once the structure is no longer maintained the walls will begin to turn back into earth after a few decades, especially if biodegradable burlap bags were used. If polypropylene bags are used, they will only biodegrade if exposed to the sun's ultraviolet light once the protective earthen plaster deteriorates, which it will do over time without regular recoatings. If an earthbag structure is dismantled prior to the polypropylene's disintegration, the bags as well as the earth inside can be reused for a new building.

TLHOLEGO LEARNING CENTRE, SOUTH AFRICA

This prototype was designed by Joseph F. Kennedy and constructed using burlap sacks and soil-cement plaster, a concrete bond beam, and a brick dome roof.

The walls were built using burlap sacks filled with the earth from the site, which had a high clay content. These were well tamped to achieve compaction. The bond beam was poured on top of the wall to which the brick dome would be fixed. The first row of bricks to the brick dome was hollow, allowing the insertion of reinforcing rods and anchor bolts to anchor the dome to the bond beam. The structure has buttresses for added stability and was plastered with soil cement. Sawn bags were also used for shade on the trellis.

Detail of the earthbag house prototype showing the brick domed roof and buttresses.

External planters, also constructed using earthbags.

EARTHBAGS FOREVER

Many regions in the world, including the desert regions of North Africa, the Middle East, and the southwestern United States, do not have abundant supplies of wood, stone, or clay. In Egypt, for example, the traditional construction material was adobe made out of clay taken from the flood plains of the Nile. Now clay is in great demand, and therefore more expensive to acquire. To be able to build with ordinary sand or unstabilized earth could benefit many people.

Of course, unlike adobe or cob construction, where the earth used contains clay as a binding element, the sandbag or earthbag technique requires a source of bags, but in most places, cloth is an accessible commodity. These bags can be made out of absolutely any cloth, even old clothes cut up and modified to hold earth. The bags are only there to hold the earth in place before plaster can be applied, unless the shelter is needed only for a year or two, in which case the bags can be left exposed and allowed to deteriorate in the sun.

Another great advantage is that sandbags have long been used to control floods and erosion in many areas. This demonstrates the strength and durability of these walls, suggesting that this material would be perfect for flood areas and disaster relief. The aerodynamic shape of domed houses that integrate into the landscape, shaped like mounds or hills, might better withstand strong winds and hurricanes, providing another advantage of these inexpensive shelters.

Ultimately, my love for the earthbag technique also comes from the simplicity of the construction process. There is no saw or nails in sight, just bags and earth. There is no need to lift heavy loads, because the earth is carried to fill the bags in place. Any child or adult could build themselves a house!

There are no straight lines in the universe, all are curved.
—Albert Einstein

2

Using Basic Structures from Nature to Build with Earth

To be able to build out of earth alone, we must understand certain basic structural principles. To me, the most important structural element, the key to all earth construction, is the *arch*. This was one of my earliest and most exciting discoveries concerning earth architecture, as the arch is a form that occurs all around us in nature, which allows us to build strong, resilient structures without high-embodied-energy materials such as timber, concrete, or steel. An arch can be used to form the curved or pointed upper end of an opening (as in a window or door) or a support (as in a truss or bridge); this shape is strong and stable because gravity pulls equally on each part, and each part supports the weight of the parts above.

Most naturally occurring caves incorporate the arch in their structure. Also, animals that make dwellings in the earth use this form to prevent their homes from collapsing. One such example is a colony of termites building their earth house using the arch as their main structural element to support a network of ducts and cavities, as described by Lewis Thomas in *The Lives of a Cell*:

When you consider the size of an individual termite, photographed standing alongside his nest, he ranks with the New Yorker and shows a better sense of organization then a resident of Los Angeles. Some of the mound nests of *Macrotermes bellicosus* in Africa, measure twelve feet high and a hundred feet across, [and] contain several millions of termites. . . . The interiors of the nests are like a three-dimensional maze, intricate arrangements of spiraling galleries, corridors, and arched vaults, ventilated and air conditioned. . . . The fundamental structural unit, on which the whole design is based is the arch.

As Nader Khalili has noted in *Ceramic Houses,* forms in nature, whether constructed or created by natural forces, exemplify efficiency.

Nature generates structures based on the principle of minimum material, maximum efficiency. From molecules, to soap bubbles . . . all follow

Facing page: The first earthbag dome attempted at Cal-Earth. Strapped straw-filled bags form the upper part of the dome.

Termites building an arch. Even though they work on the opposite ends, the arch meets in the center. (After Woodward 1995.)

WASIM MADBOLLY

An adobe dome with a cooling tower in Egypt.

Sean relaxing in the opening of Allegra's earthbag garden wall after a hard day of plastering.

Large-span concrete arch, Arcosanti, Arizona.

this general rule . . . a spider's web is a natural structure that works by ultimate tension, and an eggshell is a structure that works by ultimate compression. Both use the minimum and the appropriate material with maximum efficiency.

Over the course of architectural history, the construction of roofs and openings out of earth alone became a necessity in many regions where structural timber was not easily available. In Egypt and the Middle East, among many other places, builders came up with the idea (no doubt emulating nature) of arched openings and vaulted roofs to cover the spaces they wanted to inhabit. One can still visit large spanned domes (a set of arches with a common central peak or pivot) and vaults (a series of adjacent arches) in the Middle East built with adobe that have lasted for centuries. However, instead of building large spaces and then subdividing them, it is often more appropriate in earth construction to build several smaller spaces connected to each other, in order to provide for structural stability in each element, and to accommodate different functions, like rooms in a conventional Western building.

This introduction to nature-based architectural forms will emphasize the primacy of the arch, that singular structural element that enables builders to construct arched openings, vaults, and domes. Once you understand the structural principles of the arch, you can create a network of variously sized and shaped arches constellated together to form a variety of useful and beautiful spatial structures. When the arched form is repeated in linear fashion, it becomes a *vault*. When an arch is rotated on its central axis or centerpoint, it becomes a *dome*. When an arch is laid upon the ground, then partially raised, diagonal to the ground, it becomes an *apse*.

ARCHES

A chain hanging between two posts creates an arch form that is perfectly in tension, because of equal distribution of gravity along the curves. An arch of this shape is called a *lancet* or *catenary arch*. When you turn this shape upside down, you get a design that provides ideal compression, evenly distributing the downward, compressive forces along the whole of the arch.

A chain under tension. Chain turned upside down forms a lancet arch under compression.

In a masonry arch using this form, the individual bricks are tilted upward at their outer edges, toward the center of the arch. Once a keystone is placed in position at the peak of the arch, vertical force, or gravity, pushes down, causing the stones to press against each other and transferring the load to the ground.

To prevent this downward force from causing the sides of the arch to kick out horizontally, thereby collapsing the arch, buttressing must be added at either side around the base. The point of greatest

horizontal pressure is the point along the arm of the arch where it begins to curve toward the center to meet its symmetrical counterpart, the other arm of the arch. The imaginary line running between these two points where the curves commence is called the *spring line*. Buttresses should reinforce the base of the arch up to this spring line or higher.

You can calculate the size of the buttress needed by drawing a model to scale on paper (see the illustration below). Divide a curved arch (which could also represent a vault as well as a dome) into three equal tangential parts. Project out from Y, using point X as the center line of a circle and Y as the end of the radius. Measure from point Z to determine the necessary thickness of the buttress.

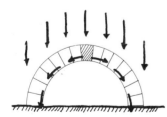

Vertical forces acting on an arch.

Horizontal forces acting on an arch.

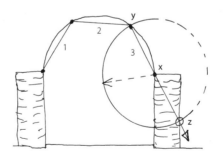

How to determine the minimum size of a buttress.

Because successive arches function as buttressing, in older buildings the arch is often repeated to form vaults, serving as a structural and decorative element, as can be seen in old churches and cathedrals. Other commonly used buttresses are *solid buttresses, parapet-tie wall buttresses,* and

Another arch acting as a form of buttress.

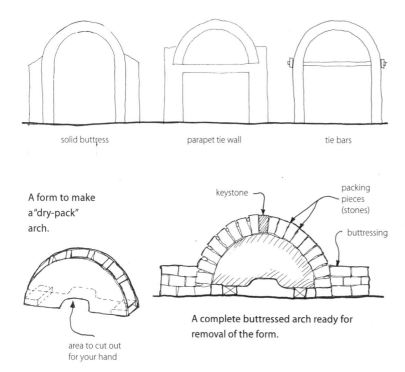

solid buttress · parapet tie wall · tie bars

A form to make a "dry-pack" arch.

area to cut out for your hand

keystone

packing pieces (stones)

buttressing

A complete buttressed arch ready for removal of the form.

Construction of a dry-pack arch at Cal-Earth during construction of the Hesperia Nature Museum.

tie bars. These buttressing systems must be evenly distributed along the vault and securely anchored in the ground.

MAKING THE ARCH

When learning about the forces of the arch and the materials that you are working with, there is no substitute for actually learning through your hands. You can make a small *dry-pack arch* as an experimental exercise. This technique is called "dry pack" because no mortar is necessary to hold the arch together. A temporary supporting form can be made out of plywood, a bucket, or anything else that has a circular shape. Balance this form on wedges to ease its removal once the arch is constructed.

As you position stones or bricks on top of the form, place shims (small stones or fired brick) between the outside edges to ensure that the inside edge of each brick is perpendicular to the arch form. Pour sand between the bricks to fill any cavities.

Once the keystone is in position at the apex or meeting point on top, and the arch is buttressed, the temporary form can be removed.

VAULTS

If you elongate an arch or repeat it in a linear sequence, the new form created is a *vault.* The same buttressing rules apply to vaults as to arches (see page 23). A vault is simply a deep or extended arch. Vaults can be used to form the passageways connecting adjacent domed or vaulted structures.

In Iran, where earthen architecture is an ancient and ongoing tradition, the most commonly found span (or width) of vault is approximately 12 feet (3.6 meters). According to research conducted by the University of Baja California, Mexico, the ratio of span to length in a vault should be no

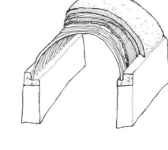

Detail of the permanent form used in Obregon, Mexico.

A repeated arch forms a vault.

more than 1.5 meters times the width. If a vault is greater in length, there is a danger that in an earthquake the vault will resonate beyond its capacity to absorb the shock, and will shatter (Khalili 1986, 57). Because the walls of a vault are structural, openings such as windows and doors should be kept to a minimum.

There are three main methods for constructing a vault.

The first is to build over a form, which can be removed and reused. This is economical for small spans, but the cost of timber to construct the form for a larger span may be considerable, unless you have a plentiful local source.

The second method is to use a permanent form, built to become an integral part of the structure. A good example of this is in the Obregon project in Mexico (see

above). This form was constructed out of a bamboolike reed called carrizo, which was bent over a very simple temporary form and buttressed at the ends with a concrete beam. Three layers of carrizo were placed over each other and were finished with an insulative straw-clay mix, capped with a waterproof coating of lime.

Another example that I have come across of a permanent form for vault construction was in the Hermosillo project (described in detail on page 143). Reinforcement rods were embedded in the ground at 18-inch (450 millimeter) intervals, then bent into position, and horizontal members were added. An expomat mesh—a stretchy metal mesh material that can be used on corners, damp-proof membranes, or any areas where plaster must be applied to a nonstick surface—was then fixed to the underside to hold the 4 inches (100 millimeters) of soil-lime or soil-cement that was applied on top. This formed a very sturdy structure. The photographs on page 146 show the strength of the arch. The reinforcement bars alone, when bent in an arched shape, could support the weight of a person. Of course, the materials were not entirely ecological.

The third method of making a vault is to use no fabricated form, merely earth. The Nubian vault may be built without any structural members or formwork, just using earth-clay-straw blocks or masonry (Houben & Guilland 1994).

The end wall is built up first. This wall is either straight or arched. The first brick is laid at an angle, and others follow suit. The vault can be started at either or both ends simultaneously to meet in the middle. For an in-depth account of this method, see Nader Khalili's *Ceramic Houses*.

Apses

An *apse* is a leaning arch. Figuratively, if you lay an arch on the ground and raise it at an angle, it becomes an apse. Old cathedrals utilized the apse form for rounded extensions to the central structure, used as more intimate chapels or sanctuaries.

An apse.

Domes

A *dome* is an arch that has been rotated on its central axis to create a group of arches with a common peak or center point.

A rotated arch is a dome.

As already noted, because of the horizontal forces tending to push the base of the dome outward, a buttress or continuous *tension ring* is necessary at the base of a dome. If an opening is made at the top, the horizontal forces will be pressing inward, therefore requiring a *compression ring,* to prevent the structure from compressing, or caving in.

As with arches, the buttressing for a dome only needs to provide support up to the spring line. The buttress can either be constructed along the outside, or the dome can be sunken into the ground so that the ground itself will act as the buttress.

A tension ring.   A compression ring.

Note: A tension ring can be created out of concrete with continuous reinforcement (rebar) at the base of the dome. A compression ring at the top is necessary only if there is an opening.

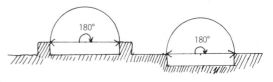

The ground acting as a buttress and a buttressed dome.

Corbeling

An alternative to the dry-pack arch discussed on page 24 is the *corbeled arch.* Corbeling involves constructing the arch in such a way that the units (bricks, stones, earthbags) lay flat, but each is stepped inward so that the weight is evenly distributed along the arc of the arch, as shown opposite.

Like the corbeled arch, corbeled domes are erected by building inward on succes-

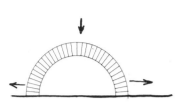

An angled masonry arch or dome.

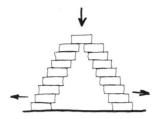

A corbeled arch or dome.

A spiral corbeled brick dome under construction at Cal-Earth under Nader Khalili's supervision.

sive horizontal courses. This is the principal used in sandbag construction; because the bags have no mortar to bind them and the sand is a fluid form, they cannot be placed at an angle. An earthbag dome could be constructed with the earthbags at an angle if there were a form underneath, but for a dome this could be very expensive to construct, given the amount of material required; therefore, angling is only practical when constructing arches. Earthbag domes must be corbeled.

To build an earthen dome on top of a square-shaped structure, *squinches* need to be constructed first. A squinch forms the transition from square into circle. Any shape with even sides could have a dome constructed above. A square is turned into a circle by creating four squinches in the four corners. In this way, a dome can serve as a substitute for a conventional truss or rafter structure.

Once you understand the "arch principle," there is no limit to the shapes that you can create, and there is no need to be bound by the conventions of rectangular architecture or even by symmetry.

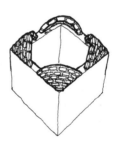

Four squinches built for each corner of a square.

Filling in the gaps between the squinches that turn the square into a circle. (Drawings by C. L. Schwarz.)

To Determine Thickness of a Shell:*

$$\frac{\text{Radius}}{\text{Thickness}} < 500$$

Radius of dome divided by the thickness of the shell should not exceed 500.

* From Philip Vittone. "Dome and Vault Engineering," *Adobe Journal* 12 & 13 (1997): 56.

3

GETTING STARTED:
DESIGN, SITING, AND FOUNDATIONS

Whatever materials are used, there is no single "right" way to design a house, as landscapes differ, climates differ, cultures differ, and the needs of residents differ. Before finalizing any decisions, research all the options that are available to you. One way of doing this is to take the conceptual design to a very detailed stage to actually gauge what materials will be needed and how the various elements will go together. Design is a process of asking yourself questions, which requires knowing what questions to ask.

DESIGN CONSIDERATIONS

Before looking carefully at site preparation and foundations specific to earthbag construction, let us consider the basic design considerations that pertain to any kind of construction, emphasizing the value of doing as much of the design as possible by yourself. By carrying out the design yourself or in close collaboration with someone who is more experienced, you can maintain control of the materials, keeping construction costs to a minimum and the complexity of the construction within the range of your own skills. You will also gain a sense of self-reliance by learning your way through the process.

Locating the Building on the Land
Whatever type of structure you want to build, however large or small, it is important to place it in the context of its surroundings so that it belongs to the land, as an integral part of the landscape. Spend as much time as possible on your intended site, preferably during all four seasons, observing where the sun rises and sets, the direction of the wind, the views. Consider the neighbors, and the simplest routes for access, snow removal, power, water, and sewage.

As much as possible, work around existing elements of the landscape such as trees and boulders. Try to minimize the damage that you will inflict on the land with even the smallest of houses. Plan to replace any vegetation that you cannot avoid destroying. Try to retain the "spirit" of the place, which is the result of this locale's unique qualities and features. You might want to leave the most distinctive areas of your land completely untouched,

Facing page: Apse and dome of the Hart's pumice-bag house in Colorado.

so they retain their natural beauty and you can enjoy them outside the sphere of your house. Good design will help you minimize your impact on the land, to satisfy both the human occupants and the wildlife, ideally enhancing rather than disrupting the natural energy flows in the surroundings, creating balance and harmony.

Landscaping

The landscaping around your house is just as important as the design of your house. Locate the house in such a way that sunlight to your flower or vegetable garden is not blocked by any part of the building. To be sure that the garden is accessible and inviting, create an easy path to it from your kitchen. If you are building with earth, you can sculpt retaining walls, benches, a bread oven, or a grotto for gathering out of the same earthen materials used to construct the house. Creating bridges between outside and inside is a crucial part of building the house.

Topography

Go to your local building department to obtain historical flood reports and other information regarding your land. Choose a higher location that is protected from runoff during heavy rains, or build appropriate drainage, contour swells, retaining walls, or gabions to rechannel runoff around the site to planted areas. Observe

Building in Flood-Prone Areas

If your land has a history of flooding and you plan to build using earthbags, you could build your house in such a way that the main living area is on a higher level, with the lower level providing a foundation of bags filled with permeable gravel or perhaps an earthbag basement with good drainage for runoff. To prevent wicking of moisture from lower to upper sections of the walls when water is standing in the lower part of the house, the walls should incorporate a moisture barrier to separate the solid-packed earthbags above from those containing gravel to facilitate drainage.

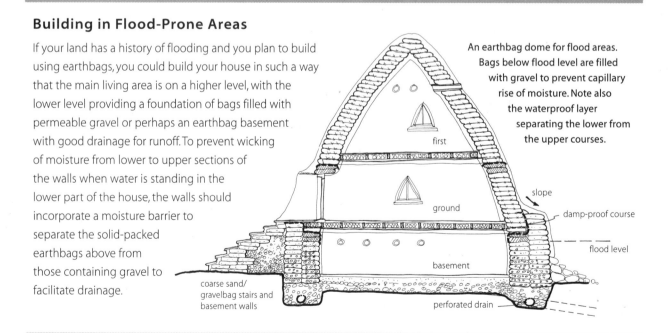

An earthbag dome for flood areas. Bags below flood level are filled with gravel to prevent capillary rise of moisture. Note also the waterproof layer separating the lower from the upper courses.

first

ground

basement

slope

damp-proof course

flood level

perforated drain

coarse sand/ gravelbag stairs and basement walls

what happens on neighboring sites during heavy rains. Talk to your neighbors about weather patterns and their experiences on the land.

Building on a hilltop increases the potential for erosion. Consider wind exposure and the aesthetic consequence for your neighbors' view if you build on a high, exposed site. Show the same respect to your neighbors that you would like them to show for you. Remember, from nature's point of view, most of the time the best house is *no* house.

Sandbags are very often used in flooded areas, and the same technology is viable in a house, where instead of merely using the bags as a barrier to water, they can be used for constructing a flood-resistant foundation or first story (see diagrams on facing page.)

Orientation

Consider the location of all windows in relation to the sun—where it rises and sets, the winter and summer angles, and the shade cast at different times of year by surrounding hills, trees, or other existing or potential buildings. If you live in an area that is sunny but has cold winters, adopt a passive solar design to take advantage of the winter sun. Especially in cold climates, avoid building on dark, damp northern slopes. (See the bibliography for recommended books explaining the principles of passive solar siting, design, and materials.)

Use seasonal screens to create covered patios in the summer that will be exposed in winter to let the sun's rays reach your home. Deciduous trees planted on the east, south, and west will provide shade in the summer but drop their leaves in the winter, allowing more sunlight to reach the house.

Carefully consider the materials of the building's interior. Earth or stone walls on which the sun's rays fall can absorb and store heat in order to give it back at night. Such walls have *thermal mass*, which evens out temperature fluctuations, retaining the night's cool to keep rooms cooler during the following day, and retaining daytime heat to keep a home warmer even on cool nights. Earthen walls can be beneficial in either hot or cold climates.

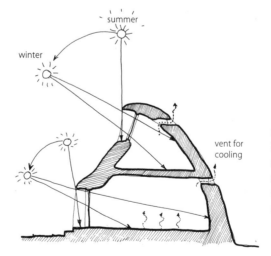

Orientation of the house toward the south provides passive solar heating in the winter. Overhangs limit solar gain in summer.

If you live in a cooler area, you will want to shield your house from cold winds. This can be done by positioning the building below a hilltop or behind dense trees and by landscaping so that foliage or other barriers provide protection. Locating a minimum of windows on the windward side can also help considerably during those windy winter days.

If you are building in a hot and humid climate, it is important to position the house in the path of the prevailing breezes. If heavy trees block the wind's path, it is good to raise the house and expose it to the breeze, with openings that go right through the house (see Steve and Carol's house, profiled on page 131). In the hot and dry climates of the Middle East, cooling traditionally has been achieved by constructing cooling towers that direct moving air right into the living spaces. Also plan for overhangs to prevent too much sunlight entering the house.

Using passive solar heating to move air through cooling towers.

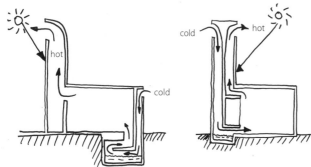

Utilities

At the earliest stages of planning, identify the location of any existing utilities, and consider what services you may need in the future, including a spring or well, cisterns for harvesting rainwater, and graywater and septic systems for processing wastes. In many countries, it is now viable to live "off the grid," providing your own electricity, heating and cooling, and hot water with solar or wind energy. With access to a year-round stream, you may even be able to harvest hydroelectricity with a microturbine sized for household needs.

Even in locations where the winters are cold and long and hours of daylight are short, heat that comes free from the sun can result in significant cost and energy savings. As emphasized above, a well-designed passive solar house can be heated almost entirely by solar gain with a small wood-burning or fossil-fuel heater as a backup.

Building Shape

Think of your house not just as a shell of a box or a dome, with all of its useful features on the inside. Remember that highly functional external features such as walls, porches, and benches provide sun and shade zones in summer and winter. If your finish will be prone to erosion from weather, elements such as landscaping, overhangs, and seasonal rain- and windscreens are essential, as well as splash-back protection at the base of walls.

Should the house be one large structure or a cluster of smaller ones either built simultaneously or added on as the need arises? A house does not have to be one structure. You could start off with the minimum living space and utilities and gradually add rooms, guest houses, and workshops. As the family grows or your needs change, more rooms or structures can be incorporated.

Think of all the activities that might be performed inside and outside the house during all twenty-four hours of the day, from when you get up in the morning until the next morning. Try to think of all the seasonal changes in light, temperature, smells, insects, the rain and snow, and the views. The more time you spend on the site before building, the easier it will be to anticipate its changes. Effective design process involves imagining the countless functional and aesthetic possibilities as well as the limitations of the materials you will be working with.

Planning Ahead

Position your house to accommodate future additions, including outbuildings, without drastic modifications of the landscape. Also consider the ways that future changes made by neighbors may affect you over time. If there aren't any neighbors, plan for the worst (a future neighbor might build a high building right against your boundary), and you are less likely to be disappointed.

As much as possible, it is helpful to identify and plan your prospective extensions of the house during the design stage to prevent awkward redesigning and reconstruction in the future. Anticipate the need for additional services and openings. If you've planned ahead well, as a family grows and more rooms are required, these can be added with minimum cost, so do not be afraid to start small and expand later, if necessary.

Site Preparation: Setting Out

Precision in the laying out of a building's base is most important for locating the foundation in the best possible place. If the earthbag foundation stem wall will be short and another building technique will be used on top of this—for example, a straw bale wall, a timber frame, or a roof type such as a vaulted, flat, or pitched roof that relies on a level surface—it's essential that the foundation provide a level plane. If you are building a monolithic earthbag structure, the foundation will be an integral part of the walls and roof; therefore the shape and size of the foundation will follow through the whole structure.

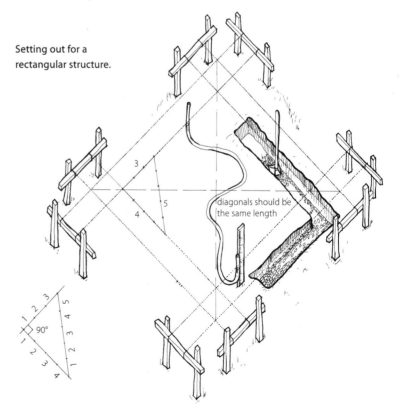

Setting out for a rectangular structure.

diagonals should be the same length

90°

The site should be cleared and leveled prior to setting out lines for the foundation and walls. If a conventional rectangular building is to be constructed, set out string lines in the traditional manner (see the diagram on page 33); at the corner points where strings cross, hammer a stake into the ground. If the strings are placed level with each other at a specific height, they can be used as a benchmark to measure the height of the stem wall or the depth of the foundation trench.

If you are building on a sloping site, the strings should be at the same height as the top of the stem wall.

A circular building or a dome requires a *compass* to begin construction. String lines are not appropriate for this shape. The compass will be useful in "setting out" or inscribing the dome, and throughout the dome's construction. The compass can be very simple, merely a string or a chain, or more complex to serve as a guide throughout the process, allowing the builders to maintain symmetry as the height of the structure rises. A more detailed explanation of how to use a compass to maintain the shape in dome construction will be found in chapter 4, "Building with Earthbags." If organic material is removed, all of the subsoil taken from the foundation area can be saved to fill bags.

Foundations

The functions of the foundation in any building are to minimize any movement of the ground over time; to spread the load of walls and roof evenly in order to give the

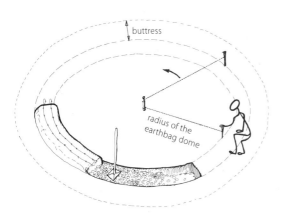

Setting out a round base house using a string as a compass.

A compass designed by Nader Khalili to construct caternary-shaped earthbag domes for the Hesperia Nature Museum.

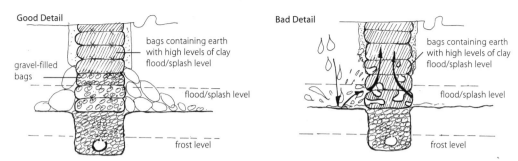

Two scenarios, with good and bad foundation details. On the left, lower bags are filled with gravel, which drains readily and prevents moisture from deteriorating the high-clay-content wall above. On the right, clay in the lower courses is worn away by rainsplash, forming cavities.

building a stable base; to hold the building in one integral unit, especially in earthquake areas; and to keep the building dry, providing a barrier between the walls and any ground moisture.

The connection of the wall to the ground is one of the most important details on an earthen house. When constructed poorly, the wall may not last as long and will jeopardize the rest of the building. If the foundation is not built carefully, moisture will migrate up the wall through capillary action and weaken the earthen walls and therefore the entire structure.

When digging the foundation trench, it is necessary to go down to undisturbed ground, below the frost heave level to bedrock or compressed subsoil, to minimize any movement caused by the ground. Once this solid ground is reached, you can build the foundation using gravel in a trench or in the bags, to raise the building's base above ground level, and to prevent capillary moisture movement. In most situations, the lower courses of earthbags should be filled with gravel, up to at least 12 inches (30 cm) above ground level, with the upper course very level to receive further courses filled with earth or any other material.

When constructing an earthbag building in a dry area, not prone to excessive moisture or flooding, it is best to build it sunken into the ground. This has several advantages:

- the earth that is dug out can be used for filling the bags;
- the ground acts as a buttress;
- the building will sit much lower on its site, so it will be less obtrusive visually, and less exposed to severe winds and weather.

When building above ground level, you will need to provide separate buttressing.

In damp areas, if the fill material for the wall construction is high in clay content, it needs a foundation base with either gravel

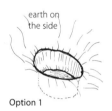

earth on the side

Option 1

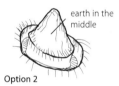

earth in the middle

Option 2

In a dry location, buttressing is provided by digging a hole so the building is sunken into the ground. Excavated earth can be used for fill.

Foundation Details

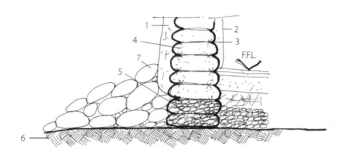

Gravelbag foundation for dry areas.

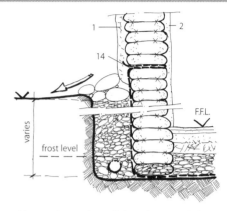

Earthbag and gravel trench foundation with damp-proof membrane for slightly damp areas.

Gravelbag foundation for slightly damp areas, without the use of damp-proof membrane.

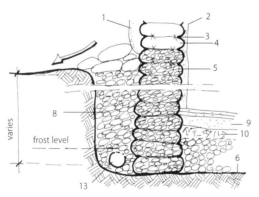

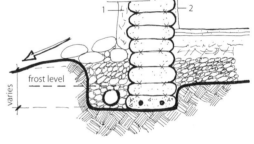

Earthbag foundation with reinforced tension ring, for domes in earthquake regions. Gravel trench is needed for wet areas.

1. Exterior finish: earthen plaster, in damp climates capped with lime.

2. Interior earthen plaster.

3. Four-point barbed wire or branches of a thorny plant. This creates friction and therefore acts as a Velcro-type of mortar between the bags. *Note:* For added stabilization of straight-wall construction in earthquake areas, the courses of bags can be pinned to each other or can be buttressed or sandwiched between wooden or bamboo posts tied to both the foundation below and the bond beam above.

4. Bags filled with earth from the site, well tamped. If the soil contains much clay, it needs to be more solidly compacted to minimize its "thirst" for moisture. Compaction of the clay reduces its ability to draw in moisture and consequently expand.

5. Bags filled with gravel (well tamped) to raise the structure off the ground, to minimize moisture migration upward into the wall through capillary action.

6. Waterproofing: can consist of a layer of clay or any other waterproofing membrane, or just well-compacted earth in very dry areas.

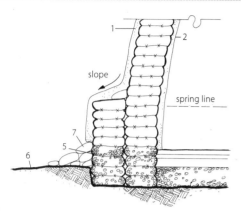

Gravelbag foundation and buttress for domes in fairly dry areas with sandy soils and good drainage. (Otherwise use perforated drain in gravel trench.)

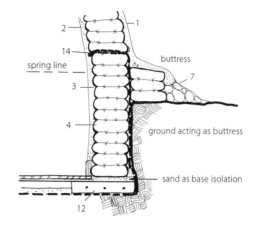

Earthbag foundation for well-drained areas with base isolation for domes and a solid "pad" or "raft" with tension ring reinforcement. Pad can be pumice-crete, lime, or reinforced concrete.

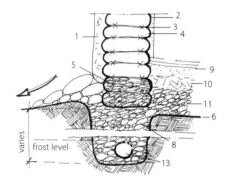

Gravel trench and gravelbag foundation.

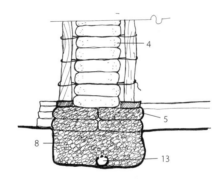

Wider foundation for straight-walled house with timber posts tied together with polypropylene or wire.

7. Large stones used for a plaster stop.

8. Well-consolidated, washed gravel in trench, to prevent capillary rise of moisture.

9. Well-compacted earthen layer. Depth varies according to requirements (this layer can contain radiant floor heating, but needs to be the necessary thickness for effective thermal mass).

10. Insulation: straw-clay mixture, pumice, or perlite-clay mixture; depth varies according to climate.

11. Well-consolidated gravel, to minimize moisture migration upward. Can be larger stones topped with small gravel. The best gravel is rounded to increase drainage spaces around the stones.

12. Pumice- or cement-stabilized lower course of bags, with a continuous ring of reinforcement in earthquake areas.

13. Drain to collect any water that may be trapped (to be approximately $1\frac{1}{2}$ inches (4 centimeters) off the base of the trench to reduce blockage.

14. Waterproof membrane.

or "sandy" fill. Otherwise, if the bags containing clay are exposed to water from flooding or even "splash back," the clay can expand and break apart the wall or dissolve and seep out, leaving cavities and resulting in instability. Alternatively, line the foundation trench with a plastic sheet to serve as a damp-proof membrane that will prevent moisture in the ground from migrating upwards into the wall above. Or wrap each bag in the lower courses individually in a plastic bag before tamping it into place (see the description of the Kaki Hunter and Doni Kiffmeyer project in chapter 8). If the area is not so damp, gravel in the trench and gravel in the lower courses of bags will facilitate drainage, minimizing the movement of moisture upward.

If your house is in an area prone to flooding, the lower courses might require stabilization with an additive such as cement or lime. Better yet, design the house to alleviate the problem of moisture altogether. For example, you can allow the water simply to pass through the lower level. Remember that if the lower courses are filled with coarse sand and/or gravel, which drains very readily, the house is not likely to have its foundation washed out from underneath (see sidebar on pages 36–37). Be aware that many "bag" materials are subject to decay, and should not be relied upon to contain nonstabilized fill in permanently or frequently wet conditions. To secure the lower course of earthbags from decay, you may fill the bags with soil

stabilized by cement, or actually with concrete, although concrete is expensive, environmentally destructive, and its use undermines many of the advantages of using the comparably inexpensive earthbags as an alternative.

To construct a foundation with insulative properties, you can tamp the lower bags full of scoria or pumice and wrap them with a damp-proof membrane, since small-sized particles tend to absorb moisture. Alternatively, the pumice can be mixed with cement to form pumice-crete.

For earthquake areas, Nader Khalili's solution was to isolate the base of the structure by laying down a layer of sand between the foundation slab or the bedrock at the base of the building, allowing it to "float" during earthquakes, "like an upside down teacup." This would reduce the risk of breakage in the walls, as no rigid pressure points are exerted upward by the ground.

The diagrams on pages 36–37 show several examples of foundation details. The variations are innumerable, as the best solution will differ slightly for each type of earth, climate, construction material, size of structure, and budget.

Earthbag or gravelbag foundations can also be used for other natural wall systems, including straw bales, rammed earth, cordwood masonry, and adobe. Especially in earthquake regions and when combining earthbags with straw bale or cob construction, rebar or other pegs can be pounded into an earthbag stem wall for added anchorage, and rubber tubing or metal rods can be left extending from an earthbag foundation to allow for compression of the adjacent bale wall.

Retaining walls can also be built with bags, but it is important to provide good drainage behind the wall and ensure that bags are properly secured against slippage. For added stability compact the bags at a slight angle toward the earth bank. In addition, many other low-cost foundation systems can be combined with earthbag or other wall systems. Following are a few suggestions. Also, *The Last Straw* magazine published a special issue on alternative foundations (no. 16, 1996; see the resources list).

Tires rammed with earth.

Earth-Filled Tires

This rammed earth technique is similar to earthbags but uses recycled tires as the permanent forms. Soil-filled tires are stacked like giant bricks to form foundations as well as exterior and interior walls. To construct a tire foundation, dig a trench down to frost depth, and place tires, ramming then with slightly moistened earth, or dig a trench below frost level, fill it with well-consolidated, washed gravel, then level the surface. Place recycled tires on the gravel and ram them full of moistened earth. Concrete may be used to fill the voids between tires. Sometimes a concrete sill is poured into forms on top, with metal anchor bolts embedded for fastening down the base of the wall above. This method of attachment is not considered adequate in earthquake regions.

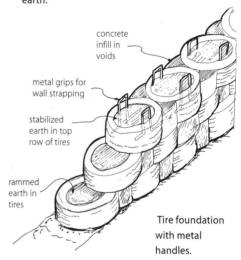

concrete infill in voids

metal grips for wall strapping

stabilized earth in top row of tires

rammed earth in tires

Tire foundation with metal handles.

Rubble or Mortared Stone

This type of foundation can be made from large pieces of stone or concrete rubble recycled from old pavement. A trench is dug, then large rubble pieces are carefully laid on undisturbed ground, with bent metal rods protruding to provide attachment points between the wall and foundation. A rubble trench can provide good drainage if necessary.

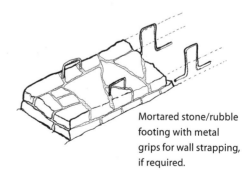

Mortared stone/rubble footing with metal grips for wall strapping, if required.

Gabions

A gabion is a latticework container woven out of willow or galvanized steel and filled with loose stones, often used as a retaining wall. Like a rubble trench foundation, a gabion will drain away moisture very effectively, so it can be used below ground as a foundation or partly above ground as a stem wall, isolating an earthbag wall from the moist ground.

Dry-stone

This type of foundation involves great artistry and requires a generous supply of relatively flat stones. Many traditional buildings in stony locales have foundations of this type. Carefully selected stones are stacked on top of each other in overlapping courses, resting on undisturbed or well-tamped earth that is below the frost line or any ground movement.

Pumice-crete

This building technique invented by Tom Watson has spread rapidly in areas where pumice is a readily available resource. A very porous volcanic stone, pumice can be used as the aggregate with a mix of a little Portland cement and water to bind it together. A typical ratio is approximately 1 part cement to between 9 and 12 parts pumice, but test samples always need to be made. For greater strength, for example above doors and windows, use 1 part cement to 4 parts pumice. Pumice-crete actually uses very little cement compared with conventional concrete, as the more finely ground pumice combines easily with cement and adds to its binding strength. Pumice-crete can be mixed and poured into temporary or permanent forms. Due to its porosity, it acts as a good insulative material, needing no further insulation, and also provides thermal mass.

Rigid temporary forms could be built in the same way as the formwork used for concrete and removed several days after pouring when the mixture has dried. Permanent forms could be galvanized wire, or any type of bags, paper or plastic, that will contain

Dry-stone wall.

Pumice-crete stem wall.

the mixture until it sets. No additional compaction is needed.

Due to its low cost, high speed of construction, and insulative and thermal properties, creating very comfortable living conditions, pumice-crete makes a good foundation, and some people are constructing entire buildings of this material. The walls of a pumice-crete house can be rendered (plastered) or unrendered. When used for wall construction, the pumice can be mixed with lime or clay instead of cement, but this will require a separate foundation to raise the wall above ground level or a damp-proof membrane to prevent moisture wicking up into the permeable wall, which could be damaged by water dissolving the lime or clay.

4

BUILDING WITH EARTHBAGS

The use of the soil-filled sacks called "earthbags" has in recent years been revived as a building technique, largely due to the pioneering work of Nader Khalili at the California Institute of Earth Art and Architecture (Cal-Earth). Its popularity is rising for several reasons.

It is low cost in terms of tools and materials, utilizing available soil in almost any region, and requiring only a few skills that are easy to learn. The polypropylene or burlap (hessian) bags used to contain the soil can be obtained free or relatively cheaply. Earthbag walls go up faster than cob or adobe and are very flexible, unlike rammed earth, allowing the construction of any shape from very straight and square structures to free forms and domes. Earthbag structures can be adapted to any conditions, from regions that flood to the most desertlike lands. When constructed properly, they are strong and durable, expected to last for hundreds of years. And, because the bags are light and easily transported, they are extremely useful for emergency shelter, in areas that are prone to flooding, or in remote locations where little or no wood, stone, or clay is available.

In chapter 2, we looked at the way that builders using masonry materials can build an arch by angling the bricks or blocks up at the outer edge, since they can be held in place by mortar, which dries solid. An earthbag dome cannot be constructed in the same way as a masonry dome. Due to the fluid properties of earth, each row of earthbags needs to be laid flat, then corbeled, or stepped inward with each successive course, in the same way that a dry-block dome is corbeled. This corbeling makes an earthbag dome much steeper than a masonry dome. The corbeled earthen dome

A brick dry-stacked corbeled dome.

Facing page: The art of earthbag engineering. Steve Kemble in the Bahamas in 1998.

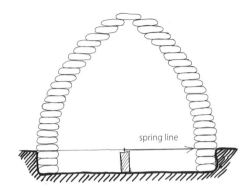

Corbeled earthbag dome buttressed by the ground.

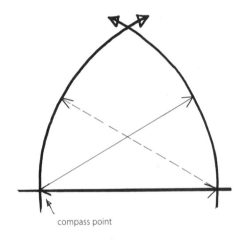

How to construct an arch of an earthbag dome on paper.

Sculpture made of burlap bags filled with sand.

takes the form of a lancet arch, as described in chapter 2. If the rows of bags have been stepped in too fast, the dome will be shallower in its rise and there could be a danger of collapse.

In this chapter, we will look carefully at the steps involved in constructing an earthbag structure. Before describing techniques for filling and tamping the bags that serve as the "building blocks" in this construction system, I will enumerate the recommended materials and tools, then review key structural principles including the use of tension rings and compression rings, as well as the importance of corbeling in a domed structure and buttressing in a straight-walled structure.

Materials

The materials needed for earthbag construction are relatively inexpensive, very portable, and available nearly everywhere. If you live in a place where these materials are difficult to find, see the resource list for suppliers who will allow you to order them for shipment to your location or use the Yellow Pages to find local suppliers.

Bags or tubes: The purpose of the bag is to retain the earth during the construction process. It is a type of permanent form to allow the earth to be placed in a course and tamped solid. When the building is subsequently plastered over these bags will no longer be visible, and will be largely redundant in terms of their structural function, since the plaster "skin" will contain the earthen walls. Again, as a general rule, the weaker the mix, the stronger the bag should be.

Bags can come ready-made or can be bought on a roll and cut to the desired length on-site (longer bags are called tubes).

Two types of bags are available on the market: burlap (hessian) and polypropylene, in a range of widths. Both types can be available in tube form on a roll (usually 1,000 or 2,000 yards per roll), or already cut up and sewn into bags. Presewn bags are generally more expensive, unless you can find a source for recycled grain, seed, or coffee bags or "seconds" with some kind of insignificant flaw that a manufacturer may be willing to sell cheap.

The width of the earth-filled bag or tube after tamping will be approximately 13 percent smaller than the width of an unfilled bag or tube, and the depth when filled and tamped will be about 30 percent of the original bag width.

Burlap is a natural woven fabric that is biodegradable, and therefore more appealing to those consciously trying to use environmentally benign materials. However, you can only use burlap if the earth is not pure sand (which will slip through the weave of the fabric) but contains compressible particles of soil. Burlap bags are heavier and bulkier than bags made of plastic, and more expensive to ship. In England, an 18-inch-wide tube of hessian may be purchased for anywhere from 30 to 50 pence per meter, plus the delivery fee. The material is delivered in rolls of a few hundred meters, and may be any desired width. In the United States, 18-inch burlap costs approximately 50 to 80 cents per yard, plus the delivery charge. For an odorless, nontoxic material, make sure that you get hydrocarbon-free burlap bags (Hunter & Kiffmeyer 2000).

Polypropylene is made of woven threads of plastic (Richardson and Lokensgard 1989). Polypropylene is a simple plastic and is not as environmentally toxic as the infamous polyvinyl chloride (PVC). It is not biodegradable, although polypropylene bags deteriorate if exposed to ultraviolet rays, so care should be taken when storing the material to protect it from direct sunlight. If a building project is not complete within three months, all exposed polypropylene bags should be covered in some type of finish to protect them from ultraviolet rays.

In England, an 18-inch-wide tube of polypropylene costs anywhere from 17 to 40 pence per meter, plus delivery fee. In the United States, most manufacturers will deliver a minimum of 1,000 yards at a cost of approximately 22 cents per yard, plus delivery fee. See the resources list for information on ordering polypropylene tubes on a roll.

There are cheaper alternatives, if you are prepared to be resourceful and persistent in tracking down supplies. It is possible to use recycled bags, which might be obtained from stores or factories that use them for bagged produce. "Misprints" are also available for a reduced price from some companies that manufacture the bags, as they sometimes make mistakes in the printing process that render the bags unsuitable to their clients. Or you can make the bags yourself by obtaining inexpensive cloth or scraps, preferably material that does not tear too easily. Fold the cloth in half and sew along one side to form a bag or tube of the desired length. If the bags are filled with a material of high binding property such as clay or stabilized soil, the bags can be removed once set.

Fill: The earth used to fill the bags can be used as it comes directly from the site, although if it contains too much organic matter or too many large stones that prevent

As a general rule, the lower the binding properties of the fill, the stronger the bag material should be.

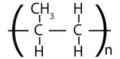

Chemical composition of polypropylene.

Polypropylene tube on a roll.

good compaction, these need to be sifted out. The soils can range from high clay content to very sandy consistency and may include other materials, such as gravel or pumice. With clay-rich soils, you could consider building instead with adobe or cob techniques. (See chapter 7 for more about clay-based building methods.) If you're determined to use earthbags, mix more sand and gravel into the mix to break up the clay, or tamp it well and ensure a high stem wall to minimize its ability to absorb moisture.

Water: This is added to the earth to facilitate the tamping, in order to achieve better compaction. The moisture content of the earth should be such that when a handful is picked up and squeezed it holds its shape, but you do not see or feel any liquid. To prevent an excess of moisture, the earth mixture can be soaked overnight.

Four-point barbed wire on a roll.

Barbed wire: This is used between courses instead of mortar to grip the bags. Four-point wire provides a good grip; as a natural alternative, you can use branches of a thorny plant or jagged rocks or stakes pounded into the bags. Barbed wire can be obtained on a coil or salvaged from an old fence. If the bags you are using are 12 inches (300 millimeters) wide, only one row of wire is needed. If the bags are 16 inches (400 millimeters) or wider, two rows may be required.

Stabilizers: These are additives mixed with the soil for increased strength, or to fortify a finish coating. Typical stabilizers are lime or cement. If constructed properly, an earthbag structure should require no stabilization. Cement can be used for bond-beams and compression rings, for extremely strong structures that must carry great loads, or for structures that are under water. Be especially careful when using cement, which while ubiquitous in our society, is associated with negative environmental impacts. Cement-based finishes should be avoided with an earthen building, because cement makes the walls more impermeable, and earthen walls must breathe over time. (The only exception is on domed structures in very wet climates.) Instead, the finishes can be earthen plasters with a lime sealant or render. Other options for finishes on earthbags are discussed in chapter 6.

TOOLS

The tools you will need to build with earthbags are simple and easy to find or make yourself.

Coffee can or shovel: Either may be used for filling the bags or tubes. Cans are easy to toss to people who are higher up on the wall, but each person will find his or her own favorite tool and technique.

Shovel for digging: A shovel with a cutting edge will make it easier to excavate soil from the site, to trench, or to collect fill.

Tamper: This essential tool is used to tamp or ram the bags flat once they have been laid in place. Garden supply stores and building centers sell manufactured tampers. You can make a metal tamper out of a piece of 1³/₁₆-inch (30 millimeter) diameter metal pipe about 40 inches (1.5 meters) long welded to a 6-by-6-inch (150-by-150-millimeter) square of metal plate about ¼ inch (6 millimeters) thick. To make an even lower-cost tamper, take a plastic yogurt cup, fill it with concrete mix, and place a stick studded with nails in the center of the wet concrete, possibly with rolled-up wire mesh for reinforcement. Let the concrete cure for at least two weeks before use.

A heavy block or chunky piece of wood can also be used to flatten the sides of an earthen wall or to beat the bags into any shape desired.

A stand: A fold-out stand will aid in the filling of small bags. (To ease the filling of longer bags or tubes, prop a cut-off piece of pipe in the opening.)

Water source, or buckets for hauling water: If the soil is too dry, water may be needed to help make an earthen mixture more compactable. One of the reasons that earthbag construction is very well-suited to dry locales is that you can also fill bags with dry sand, gravel, or sifted soil.

Water level: for leveling the ground.

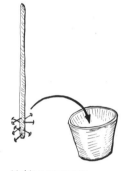

Making a tamper.

Left: The tools for earthbag construction.

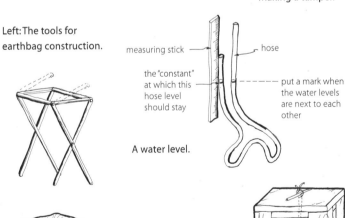

measuring stick

hose

the "constant" at which this hose level should stay

put a mark when the water levels are next to each other

A water level.

A small bag-filling stand.

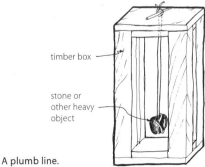

timber box

stone or other heavy object

A plumb line.

Section of earthbag dome showing the compass in use.

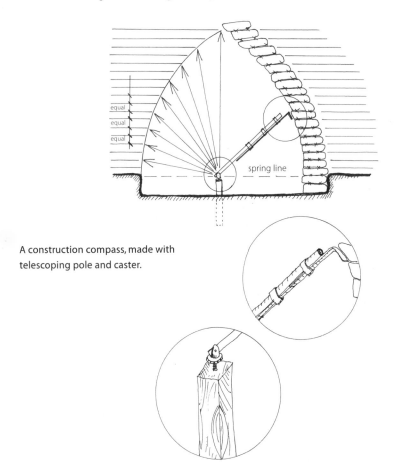

A construction compass, made with telescoping pole and caster.

Plumb line: A plumb bob and line allows you to check a straight wall for vertical levelness.

Wheelbarrow: This is used for transporting material, for mixing cement or lime into the earth to stabilize it, or for mixing adobe or cob plaster.

Hoe: This may be more practical than a shovel for mixing ingredients.

Blade or scissors: For cutting the bags or tubes.

Wire cutters, level, ladder, tape measure, gloves, and trowel: All of these will prove useful on any building project.

Compass: Required only as a placement guide for the bags in the building of symmetrical domes, a compass can be as simple as a chain or a string, or more complex. When the courses of the dome reach the stage at which they must start to curve inwards (the spring line) the compass needs to be extended after each row. This can be worked out by making a drawing of the dome, as in the drawing at left and page 44.

An extendable compass may be made with a length of hollow pipe (electrical conduit will work, or one of those telescoping poles used for cleaning swimming pools) attached at one end to a caster (such as those on a grocery cart) from which the wheel has been removed. The caster will allow for rotation as well as up and down movement. The caster should be affixed to a 4 x 4 (100 x 100 millimeter) post planted upright in the ground at the center of the dome. At the upper end of the pipe, use pipe clamps to attach a guide made of an L-shaped piece of metal.

PREPARING THE FILL

As I have explained, the bags are used as a temporary formwork for the tamped or rammed earth during construction of a building and before the plaster is applied. The plaster finish can be seen as a long-term sheathing, which if maintained attentively can be considered "permanent." The material that goes into the bags can therefore be of any consistency ranging from very loose—for example gravel, pumice, or sand—to a more compactable soil that contains varying amounts of clay.

The best fill for an earthbag wall is one with the same consistency as the traditional mix for rammed earth: approximately 25 percent clay to 75 percent sand, which will dry into a cementlike hard block. However, earth seems to be sufficiently compactable with as little as 5 percent clay.

With soil that has a high proportion of clay, it is probably better to consider an alternative type of construction, such as cob, adobe or one that makes use of other fiber/clay composites; because the clay has binding properties, a bag to contain it is not really necessary. However, bags are faster to construct than cob.

If pure sand is used, you must take several precautions. The sand *must* be made slightly damp to facilitate the compaction. The bags need to be wider to allow for more stability, more buttressing needs to be provided, and if domes are being constructed using the corbeling method, care needs to be taken that each row does not step in too fast. This type of dome can actually be much taller than masonry domes, with steeper sides and larger buttresses, such as on Shirley Tassencourt's dome (see diagram, page 124). For increased stability, it is also possible to tie down one course of bags to the two below, creating a net with wire mesh or strapping, or to construct the dome over a permanent form such as demonstrated in the Malawi project (see page 142).

Whenever earthbags are used, but especially in flood areas, care needs to be taken that the lower courses of the wall do not contain clay, and are properly detailed to shed water (see drawings on page 35). If bags containing clay are exposed to water, the clay can either expand and break apart the wall or dissolve and seep out, leaving cavities that create instability. The higher courses need to be tamped or rammed well to reduce the ability of any clay in the bags to absorb moisture.

Generally, the earth excavated on site can be used for fill. Be sure to remove topsoil and set it aside for a future garden, instead using the subsoil for construction. Remove all large stones and organic matter from the earth mix, as these materials too could create cavities later on. If a large number of stones are found in the soil, sift them out using appropriately sized screens, and use them as gravel in the lower courses of bags or for foundations, as this nonabsorbent material will drain well and prevent capillary rise of moisture. If crushed pumice or scoria (a porous volcanic stone) is used for earthbag walls, it will serve as both thermal mass and insulation (as in the Hart's house, profiled in chapter 8), but smaller-sized particles can wick moisture, so care should be taken not to use too much fine pumice or scoria in the lower courses, which are prone to being damp unless a damp-proof membrane is used, or unless the pumice is mixed with cement. (For discussion of pumice-crete, see chapter 3.)

If the earth used for filling is totally dry when dug out, it needs to be sprayed with water to ease compaction in the tamping process. More compacted fill is ultimately more stable. Remember, the soil should be moist enough that when a handful is picked up and squeezed it holds its shape, but you do not see or feel any liquid, and when a lump of the soil is dropped it falls apart. The soil can also be soaked overnight.

The appeal of the earthbag method is that builders can use such a wide variety of soil and other types of fill for construction,

only attending to these very general guidelines: not too much clay, not too many large or sharp rocks, not too dry or too wet. For information about doing soil tests when building structures that require more refined sensitivity to soil contents, see chapter 7. For information about stabilization, see chapter 6.

Filling Bags or Tubes

Before beginning to fill the bags, make sure the earth is moist enough to allow compaction. Bags may be filled in several ways: with a shovel, tin can, bucket, or whatever

Two strands of barbed wire placed between bags during construction act as a mortar.

A team of three filling a long tube in Mexico.

the person doing the work can lift. It is never necessary to lift the bag itself: the bag stays in place and earth is brought up to its opening. Bags higher on the wall can be filled in place. Cans of soil can be thrown up to the person doing the actual filling (see photo on page 53). Bags filled by different people will vary in thickness, due to differences in strength and technique. It is therefore important that one person or team builds a whole row to minimize the changes in thickness during one course. The rows may vary from each other, but as long as each row is of a consistent thickness, this is not a problem. Once a row is complete, tamp it well, then place two strands of four-point barbed wire on top as keying for the next row. As you stack bags in successive courses, remember to always stagger the joints, just as in masonry construction.

Each tube or bag can be cut to the desired length. How long do bags need to be? If you are building a dome, it is good to have one continuous length all the way around, unless this would mean a length of more than 30 feet, because beyond that, the tube becomes difficult to fill.

For the foundation and the first few rows of wall, it is good to use bags or tubes that are as long as possible to minimize breaks, for structural stability, but even small bags that are well tamped can be used for the foundations. When measuring the required length, add an extra foot of fabric at each end of the bag. You will later fold over the surplus fabric to close the bag.

A mechanical pump used for pumping concrete can also be used to speed up the process of filling the bags. For landscaping or large industrial projects, where time and labor are costly, professional builders might use a continuous berm machine, which extrudes a fabric-encapsulated continuous berm of sand, rock, or native soil at a rate of 10 to 50 feet per minute.

Filling a Bag with More than Three People
Two teams can fill a long bag or tube from the two ends using the following procedure:

1. Cut the tube to length (up to 30 feet).
2. Fold the edges of each opening back as far as possible toward the middle, as shown to the right.
3. When the folding reaches the midpoint, start filling the tube from each open end. As one person shovels in the earth, the other can hold the end of the tube open and unfold the ends as it is being filled.

The speed of this method of filling bags on the project in Mexico (see chapter 8) averaged out to be 25 feet per hour, per team.

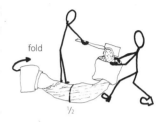

Filling the bag from both ends.

Filling the long bag from one end.

"Diddling."

Filling Bags or Tubes with Only One to Three People

When working with fewer people, the bag or tube should be filled from only one open end. Once the desired length of bag is cut, shovel or scoop as much material as you can lift into the bag, then shake it down to the opposite end. While one person shakes the earth toward the end, a second person can step on that end to prevent the material from flying out.

Or, when working with smaller, more portable bags with one end sealed, you can shake the material down into the open end, and lift the whole bag into position. Once the bag is in place, fold under the open end of the bag to close it. When a row is complete, this bag should be tamped solid and flat before the next course is placed on top and is ready for barbed wire.

Another way to fill bags with a smaller crew is to slide a cut piece of wide pipe into the open end of a bag, like an ankle in a sock. The pipe forms a chute for the earth to go through. Each time a manageable amount is shoveled in, take out the pipe and shake down the bag. Be sure to avoid letting the bag bunch up while it is being filled; each load of earth must go all the way to the end, with no gaps. A brick can be placed under the bag to prevent creases in the bag, or one of the crew can use a foot to support the lower side of the bag as it is filled. Use a piece of sheet metal or a board to protect feet and clothing from the barbed wire when working on top of a course that has been keyed with wire.

Using Small Bags

If small bags are used, the wall will tend to have a great many bag corners sticking out, which makes plastering difficult, as much more plaster is needed to cover these. To prevent this, tuck in the bottom corners as the bags are being filled. Utah earthbag builders Kaki Hunter and Doni Kiffmeyer named this process "diddling." When filled, the open end of the bags can then be gently lowered into place and those corners diddled as well, tacked under the weighted end of the full bag.

One person can fill small bags by using a stand to keep the end open. Remember, the small bags must be laid in a running bond, with all joints staggered. The time required to fill small bags for the Honey House project (see chapter 8) averaged four bags (approximately 6½ feet or 2 meters) per hour, per person.

TAMPING

As noted above, to minimize unevenness, each bag in a row should be filled to its maximum capacity by the same team. To create as level a wall as possible, do not tamp until the whole row is filled. Once the whole row is laid, it can be tamped until no movement of the earth is felt. The sound of the tamping changes as the earth in the bag is compacted, becoming less of a "thump" and more of a solid "smack."

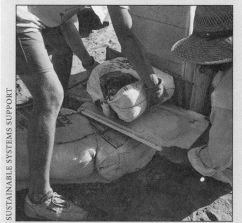

SUSTAINABLE SYSTEMS SUPPORT

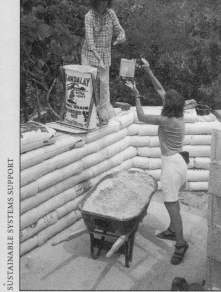

SUSTAINABLE SYSTEMS SUPPORT

Far left, top: Using the metal tray to position the first bag correctly.

Far left, bottom: Filling the long tube using a plastic pipe as the chute.

Left: Filling a bag using a stand.

Below: Steve Kemble tamping an earthbag wall with a concrete tamper in the Bahamas.

Inset: Untamped and tamped sides of a wall.

SUSTAINABLE SYSTEMS SUPPORT

SUSTAINABLE SYSTEMS SUPPORT

untamped wall   tamped wall

Buttressing wire. Each row of bags is tied to the two rows below.

Tamping soils with high clay content lessens the earth mixture's tendency to draw in moisture, but does not eliminate this tendency entirely. Be sure to avoid having too high a proportion of clay in the earthbags, especially in the foundation or lower courses, which are more likely to be exposed to water.

You may also wish to tamp the sides of the wall, checking the vertical straightness with a spirit level or plumb line. An advantage of tamping the sides is that then the wall surface will require less plaster. The disadvantage of a very level or even wall surface is that the plaster has less surface area to key into.

KEYING

After tamping, each course needs to be keyed with four-point barbed wire or branches of a thorny plant, which will provide friction to prevent any shifting of the bags over time. If no barbed wire is available, the bags can either be well buttressed, tied to the bags below (see Kelly Hart's project in chapter 8), sandwiched between wooden poles, or pinned with reinforcement rods. For very wide and short stem walls or landscaping walls, barbed-wire keying is not necessary. Also, on smaller structures, using rough rocks or chunks of gravel between courses will provide adequate keying.

If the bags used are wider than 14 inches (350 millimeters), or a dome is being constructed, two rows of keying may be necessary. This keying is especially important in corbeled domed structures, providing tensile strength while enabling each row to step in slightly.

STRUCTURAL REINFORCEMENT AND BUTTRESSING

In domes, there are two areas of maximum pressure that require careful attention, especially in areas of high winds or seismic activity. The base of a dome can be buttressed on the outside with the ground, with constructed benches, or reinforced with a "tension ring." The other point of pressure is the top of the dome. If there is any opening at the top, it must be reinforced with a "compression ring."

A tension ring is a continuous and rigid ring at the base of the

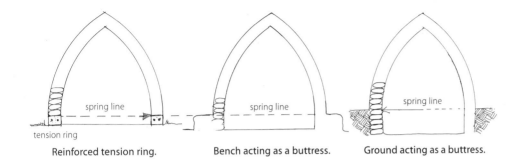

Reinforced tension ring.   Bench acting as a buttress.   Ground acting as a buttress.

dome, which absorbs the downward horizontal forces that otherwise would cause the base of the dome walls to splay out and collapse. This ring has the same function as buttressing and is needed in all structures in seismic zones and in domes that do not have some form of buttressing around the base. In most domes, unstabilized material may be used in the bags with which the walls are constructed, but in seismic regions the ring around the base must be stabilized with continous reinforcement surrounded by concrete or cement-

stabilized rammed earth, metal, or some other resilient material. Since this reinforced ring may be the most expensive element in an earthbag dome, it may be advantageous to consider buttressing the structure in areas that do not have seismic activity (see the diagram series above).

A compression ring is the tension ring's counterpart at the top of the dome, necessary if there is an opening there. It prevents the dome from caving in because of

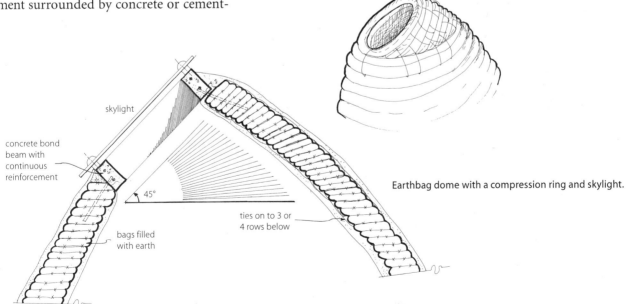

Earthbag dome with a compression ring and skylight.

Buttressing during construction at Allegra's house, Arizona.

DOMINIC HOWES

Staggered joints at the corner of the Three-Vault House, Mexico.

Long bags "woven" into place during construction of the author's retreat.

the upward, inward pressure of that opening. Like a tension ring, a compression ring needs to be continuous, made of concrete, metal, wood, or other material containing sufficiently sized continuous reinforcement. For more complex structures, consult an engineer.

While curved walls are structurally self-supporting, straight walls need additional support. The diagrams on this page show

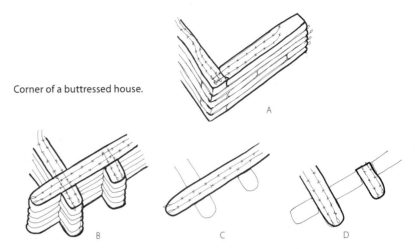

Corner of a buttressed house.

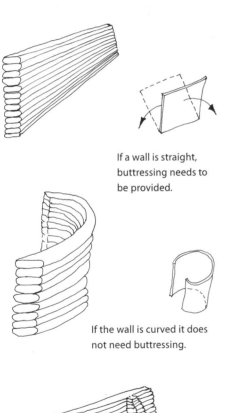

If a wall is straight, buttressing needs to be provided.

If the wall is curved it does not need buttressing.

A buttressed wall.

ways of reinforcing a straight wall with a corner or with connected buttresses. Remember, for stability when constructing intersecting walls and buttresses, always stagger all joints.

Openings

In an earthbag dome, the number of openings cannot be too many or else the structural stability of the dome will be compromised. The distance between openings should be large enough to properly buttress the arch that forms each opening. In general, openings that are square are best suited for square houses; it is possible to create small square openings in domes if there is a lintel, but structurally this is not a good idea.

During wall construction, where there will be openings, leave loops of wire extending out from the strands of barbed wire laid between the bags, which will

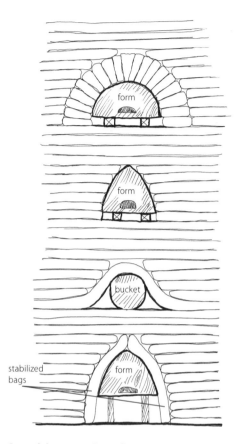

Several demonstrations of creating an opening without the use of a wooden lintel.

Shirley's dome, showing different formwork on top of the wall.

allow you to tie off these wires up and down around the opening for added strength. Some builders reinforce their openings with wire mesh, which can also be attached using these wire ends.

There are two categories of openings: *arched,* which do not need wood, metal, or concrete as a lintel above, or *square,* which need lintels, or which utilize a bond beam at the top of the wall as the lintel.

Arched Openings

If the opening is in the shape of an arch, no lintel is necessary. One way of leaving an opening in the earthbag wall is to use a form. A form can be a circular object such as a wheel, a bucket, or a barrel, or can be specially constructed out of timber in the exact space desired. The opening can also be filled with earthbags that are subsequently removed.

Several types of arches can be created, some of which are shown in the drawings and photos. To make the "curved bag" window shown above, stabilized earth must be used.

To make a form, cut an arch in the required shape out of plywood or other flat material and duplicate it, then attach these flat arches to each other with pieces of timber so that the ends are parallel. Next, cover the whole arch with plywood or other flexible, sheetlike material.

Make the form at least 2 inches (50 millimeters) wider on each side than the intended size of the finished opening, to allow for the thickness of the plaster.

When placing the form, make sure to position it on top of wedges, as this will ease the removal of the form after the arch is complete. These wedges could be chunks of timber or tapered log ends.

Place a minimum of three well-tamped courses of bags above the opening before removing the form.

When building a vault that connects to a structure through an arched opening, the main opening should be located at the end where the arched opening in the wall has no load-bearing function (see page 26). Some builders have constructed reforms for vaults,

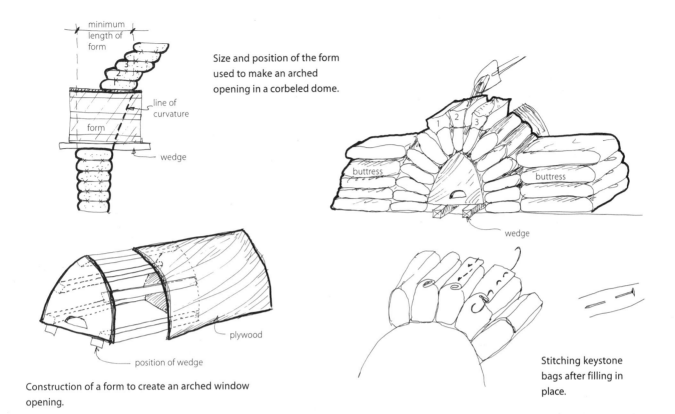

Size and position of the form used to make an arched opening in a corbeled dome.

Construction of a form to create an arched window opening.

Stitching keystone bags after filling in place.

but this can be expensive because of the quantities of materials required. It is less expensive to make a permanent vault out of bamboo or other bendable material, which the earthbags can be layered around, or a Nubian vault out of adobe (see page 25–26). If you do decide that a temporary vault form is necessary, it is most economical to make one that can be reused rather than demolished.

When the earthbag wall reaches the height of the opening's bottom sill, place a form where the opening will be, positioned on top of wedges to ease its removal after completion of the arch.

As you stack the first three rows of bags around the form, add additional permanent buttressing on either side of the primary bags to contain the horizontal forces that will act upon the finished arch.

The arch's "keystone" will consist of the last three bags laid in position. These are placed with their tops still open, then filled up with additional earth from above, as shown above right. The earth must be shoveled into the last three bags simultaneously, to create the keystone effect. To close these bags, either use nails (stuck like tailor pins through the fabric of the bags), or stitch them closed with a piece of wire.

Top: Window in the end of a vault, Mexico.
Middle: Stabilized arch, California.
Bottom: Arched forms.

As emphasized above, before removing the form, you must complete the rest of the wall with at least three tamped rows laid on top of the arch.

Square Openings

Although square openings for windows and doors are not structurally sound in dome construction, you can incorporate square openings in a straight-walled earthbag building by providing rough framing ahead of time or by cutting them out from the finished wall, provided the necessary buttressing is in place prior to this excavating.

As shown on page 61, it is important to add sturdy diagonal bracing to all window and door frames to keep them square during construction. This bracing can be removed once the walls reach full height. Door frames also need to be securely attached to the foundation for stability.

The detailing around window and door openings—for instance, the seal between a windowsill and the earthbag wall below—is extremely important in order to prevent penetration of moisture. Also note that if a timber frame is used to provide the structure for a building where earthbags serve as infill, it is important to separate wooden members from the earthwall with a carefully attached waterproofing membrane to prevent moisture penetrating through at seams between the earthbags and wooden posts and beams or framed openings. (See chapter 6 for more information on waterproofing.)

BOND BEAMS

Known in conventional construction as a "plate," a bond beam is a rigid structural unit, usually made of wood, metal, or concrete, that sits on top of a wall and evenly distributes the weight of a subsequent floor or the roof. While not relevant to dome construction, bond beams are important in straight-walled construction to tie together and stabilize the earthen structure at its point of greatest outward pressure, especially in areas of high winds or earthquakes. In addition to serving as a level platform for a roof, a bond

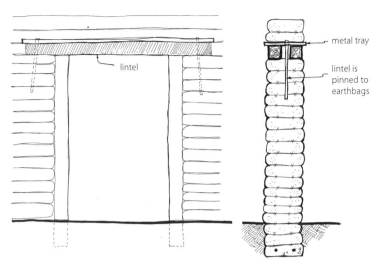

Elevation of doorway.

Section through doorway.

metal tray

lintel

lintel is pinned to earthbags

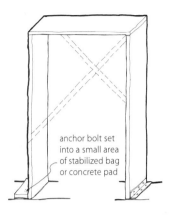

anchor bolt set into a small area of stabilized bag or concrete pad

Door frame securely fixed to stern wall with diagonal bracing.

Below and bottom right: Window details. Always slope the external window sill away from the building with an overhang of at least 2 inches (50 mm) and a drip edge.

window

timber windowsill with a drip edge

timber sill plate

metal lath

cement-stabilized earthbag

damp-proof course

internal plaster

anchor bolt to fix sill plate to cement-stabilized earthbag

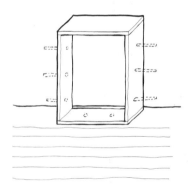

Window frame.

sculpted earthen sill with a drip edge

metal lath

window plate

timber sill plate fixed with an anchor bolt to a cement-stabilized bag

internal earthen plaster

well-tamped earthbag

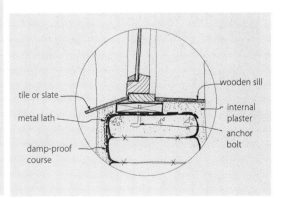

tile or slate

metal lath

damp-proof course

wooden sill

internal plaster

anchor bolt

beam can also serve the function of a lintel, spanning the unsupported gap in a wall created by a window or door.

In an earthbag building, depending on its function, the bond beam can be made of timber, steel-reinforced concrete, or even cement-stabilized earth, well tamped in the bags.

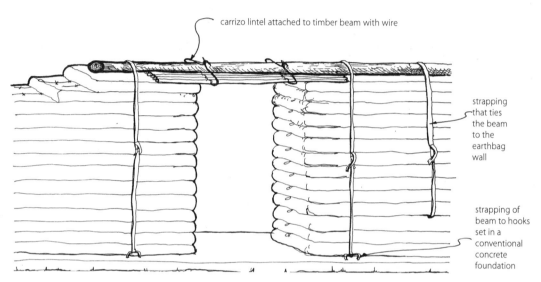

carrizo lintel attached to timber beam with wire

strapping that ties the beam to the earthbag wall

strapping of beam to hooks set in a conventional concrete foundation

Simple log bond beam and carrizo lintel.

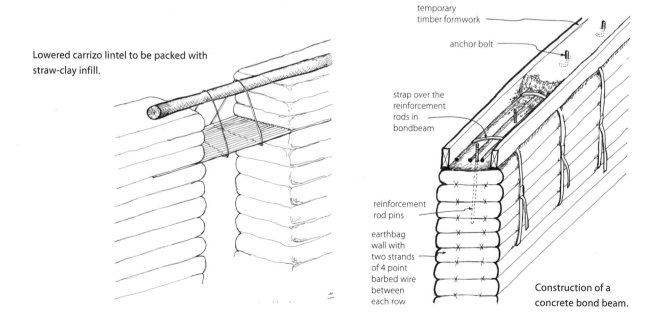

Lowered carrizo lintel to be packed with straw-clay infill.

temporary timber formwork

anchor bolt

strap over the reinforcement rods in bondbeam

reinforcement rod pins

earthbag wall with two strands of 4 point barbed wire between each row

Construction of a concrete bond beam.

SUSTAINABLE SYSTEMS SUPPORT

SUSTAINABLE SYSTEMS SUPPORT

Far left: A bond beam in the Bahamas.

Left: A formwork for the bond beam, which will be used as a base for the timber wall plate of the next level.

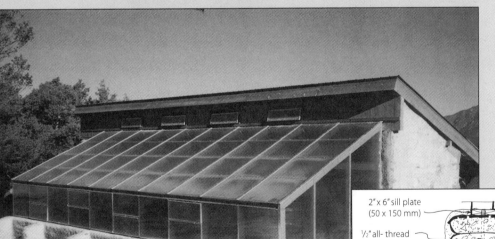

KELLY HART

Below: A conventional greenhouse in Colorado constructed over a pumice-filled bag stem wall.
Inset: Section through the stem wall of the greenhouse showing the wall plate detail.

2″ x 6″ sill plate (50 x 150 mm)

½″ all- thread

well-tamped pumice filled bags

well-consolidated gravel in trench

damp-proof course

row of well-tamped cement- stabilized soil

papercrete (fibrous cement) internal and external plaster

well-tamped gravel bags

5

ROOFS

The roof is one of the most important factors keeping a building dry and warm. A good roof protects the inhabitants from rain, snow, wind, the cold, and the heat. It will shed water away from the house, directing it at the garden, or will catch the rainfall to be stored for later use. The roof can be a dominant feature, making a house stand out, or can help a house disappear gracefully into the landscape.

The ideal roof for earthbag walls is one constructed using the same materials as the walls. The main attraction to most people who discover this building technique is the possibility of using no wood,

Facing page: Cooling tower of the three-vault house at Cal-Earth.

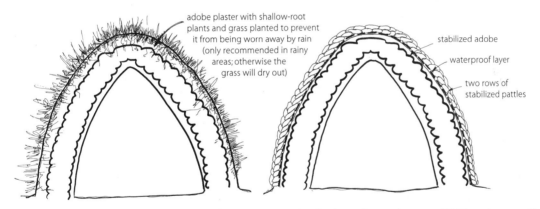

A grass covering. In rainy climates a waterproof membrane is necessary.

adobe plaster with shallow-root plants and grass planted to prevent it from being worn away by rain (only recommended in rainy areas; otherwise the grass will dry out)

Covering for damp climates (see page 98). The outer stabilized layer could also be made of papercrete.

stabilized adobe

waterproof layer

two rows of stabilized pattles

Natural roofing materials for earthbag domes.

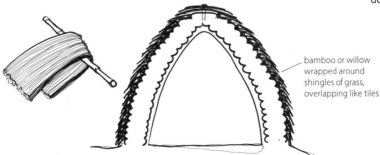

bamboo or willow wrapped around shingles of grass, overlapping like tiles

65

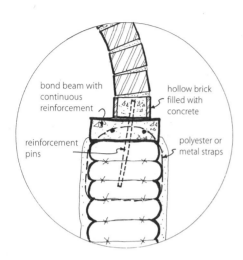

Detail of a brick dome showing the bond beam.

Surface Finishes

Surface finishes for domed earthbag roofs, to make them water-resistant though not necessarily waterproof:

- lime render or whitewash
- cement-stabilized soil
- papercrete
- earthen plaster with lime render and whitewash

In wet climates a waterproof layer is necessary underneath the plaster or painted on top.

metal, or concrete, as well as the aesthetic value of an earthbag structure. A dome, for instance, is quite an amazing space to be in (some would say "nourishing for the soul") with an option of adding at least one other floor while retaining its height and beauty. Using a dome or vault is a roof-building technique that is financially and ecologically economical if no plentiful, renewable source of wood is available. Concrete and steel contain high embodied energy, as well as being expensive, and earth is a far healthier option.

Yet, creating an earthbag dome might not be a solution to house design in all climates, nor for every type of budget, culture, or individual. The amazing aspect of the earthbag technology is that it is genuinely adaptable, allowing each individual to create a house tailored to his or her needs. In many cases it is beneficial to combine the earthbag wall system with a flat, pitched, vaulted, or other roof system, depending on the requirements. For example, in areas prone to high levels of rainfall throughout the year, it is a good idea to combine the earthbag wall system with a more conventional roof type that provides an overhang to protect the walls from the constant rain.

Brick or Adobe Roofs

An earthbag building can be covered with a shallow dome constructed out of masonry brick or adobe in areas where the climate is relatively dry.

According to an old recipe I found on the straw bale listserve (http://solstice.crest.org/efficiency/strawbale-list-archive/index.html), a cheaper and more beautiful way of waterproofing exposed bricks than covering them with cement or boards is to:

Stir 1 pound of finely powdered flowers of sulfur into 8 pounds of linseed oil. Bring the mixture to a heat of 278 degrees Fahrenheit, and then allow it to cool. Add some drying oil and paint the bricks with the compound.

For more on waterproofing, see the discussion of finishes in chapter 6.

VAULTED ROOFS

To my knowledge, a vault wider than 5 feet (1.5 meters) has not yet been successfully constructed using earthbags. Most vaulted roofs are constructed from other materials and joined to the earthbag walls using a bond beam.

If the soil contains some clay, one way of constructing a small vault out of earthbags (approximately 3 feet [1 meter] in width, which can serve as a connecting space between domes) is by standing the bags up and placing them in a leaning arch, as shown in the top diagram.

Another better way of constructing narrow vaults is to create a series of sturdy, self-supporting arches that, when joined together and finished with plaster, create a continuous vault.

Ratio of Vault Width to Length

In vaults constructed out of earth (for example, adobe), a safe ratio of width to length should be equal to no more than:

length = 1.5 meters x width

with the width not exceeding 12 feet (4 meters) overall.

For lancet vaults, the following ratio applies:

rise of vault = width/2 + 19.5 inches (50 centimeters)

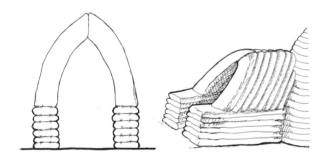

A narrow vertical leaning vault, as constructed for a passageway in the retreat.

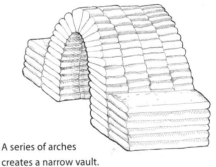

A series of arches creates a narrow vault.

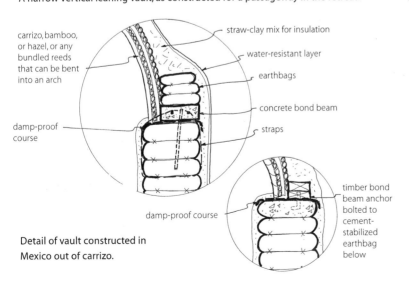

carrizo, bamboo, or hazel, or any bundled reeds that can be bent into an arch

straw-clay mix for insulation

water-resistant layer

earthbags

concrete bond beam

straps

damp-proof course

damp-proof course

timber bond beam anchor bolted to cement-stabilized earthbag below

Detail of vault constructed in Mexico out of carrizo.

Carrizo and earthen vault.

CONVENTIONAL ROOFS

More conventional roof systems can also be constructed on top of the earthbag walls. The roofs are for areas with high rainfall. It is always important for the roof to have large overhangs (at least 18 inches [45 centimeters]) on the sides, even the less exposed ones. From the standpoint of environmental impact, it is better to create roofs out of small sections of wood, if wood is necessary at all, instead of timbers from old-growth or slow-growing trees. Wood can be obtained from fast-growing trees in carefully managed forest plantations, or builders can use wood-efficient trusses, laminated timbers, or other engineered wood-fiber products. Trusses can be purchased prefabricated or can be built on-site with local materials. A typical 2 x 4 truss provides a great deal of space for insulation but does not provide extra living space, unless the scissor truss is used (see top diagram).

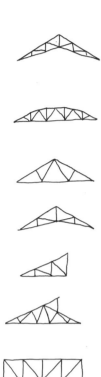

Types of trusses.

WATER-CATCHMENT ROOFS

Another supplementary purpose of a roof can be to catch rainwater to be used for washing, flushing toilets, watering the garden, and even drinking after filtration. For directing run-off to a water-storage reservoir, a good roof material is zinc-coated metal, because ethylene propylene diene monomer (EPDM)—the synthetic rubber commonly used for roofs, pond liners, and water tanks—is apparently slightly toxic. The water can be collected into a cistern located either inside the house or out in the garden. If it is placed inside the house, care needs to be taken of its location to avoid its functioning inadvertently as a very large heat sink, constantly drawing heat from the house's living space. In a passive solar building, the additional thermal mass of the water in the tank can help store solar gain. The water tank should be insulated, which can be done with straw bales.

THATCHED ROOFS

A long tradition in England, Ireland, and Wales, thatched roofs are still in use today on most cob buildings. Throughout northern Europe, thatch was made of a common reed grass (*Phragmites*) or tight bundles of straw, usually wheat or rye. Thatch conforms nicely to curved and irregular roof shapes. The biggest advantage of thatch, in addition to its aesthetic value, is that the thatch itself is the waterproofing layer and therefore does not require the

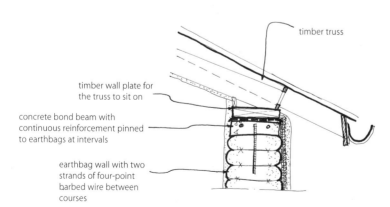

timber truss

timber wall plate for the truss to sit on

concrete bond beam with continuous reinforcement pinned to earthbags at intervals

earthbag wall with two strands of four-point barbed wire between courses

Example of how the roof sits on the bond beam.

Allegra's roof under construction, Arizona.

DOMINIC HOWES

A thatched house, Poland.

Rainwater-collecting roof, Colorado.

addition of any artificial waterproofing materials; moreover, thatch provides sufficient insulation. A well-made thatch roof can last a long time: straw thatch up to forty years, and reed up to sixty. The main disadvantage of thatch is that it is combustible, but the fire danger can be substantially reduced by incorporating measures such as ceilings that reduce airflow to the roof, a sprinkler system, or treatment of the roof with flame retardant, as discussed in Michael G. Smith's book *The Cobber's Companion* or Michel Bergeron and Paul Lacinski's book *Serious Straw Bale* (see the bibliography).

LIVING ROOFS

A "living" roof is one that supports an earthen mulch and plantings of grass, mosses, or even a berry patch. This kind of roof can be aesthetically pleasing, and can make a house blend in to its surroundings. The earth on the roof serves as extra protection for a waterproof membrane beneath, and in addition to helping the house retain its coolness in the heat of the summer, the thickness of the roof covering is a sound insulator. Such roofs have also been known to protect houses from external fires.

Straw-clay rolled onto long straw, reed, or jute spanning between rafters. When straw-clay rolls are used, no board or mat is required to support the straw-clay mix.

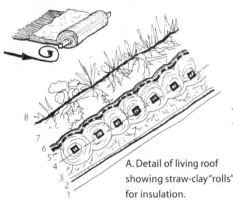

A. Detail of living roof showing straw-clay "rolls" for insulation.

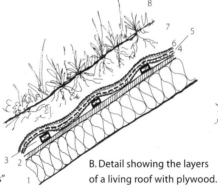

B. Detail showing the layers of a living roof with plywood.

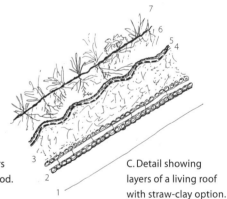

C. Detail showing layers of a living roof with straw-clay option.

1. Rafter
2. Plaster finish.
3. Straw-clay.
4. Long grasses or reeds covered with straw-clay and rolled up to create an insulating layer.
5. Two layers of waterproof membrane.
6. Corrugated cardboard or carpet scraps for cushioning to prevent puncture of the waterproof membrane and to give the roots a base to wrap around.
7. Soil.
8. Plants.

1. Insulation between rafters covered with a ceiling finish.
2. Plywood or other rigid board.
3. Timber batten (50 x 100 mm) to stop the soil from sliding off an angled roof.
4. Corrugated cardboard or carpet scraps.
5. Two layers of waterproof membrane.
6. Corrugated cardboard or carpet scraps for cushioning to prevent puncture of the waterproof membrane and to give the roots a base to wrap around.
7. Soil.
8. Plants.

1. Rafter.
2. Carrizo decking.
3. Straw-clay for insulation.
4. Two layers of waterproof membrane.
5. Corrugated cardboard or carpet scraps.
6. Soil.
7. Plants.

This type of roof needs enough rainfall to ensure the watering of the vegetation, or the roof can be planted with local plants that do not require watering. Ideally, the roof's pitch should not exceed 35 degrees, or the mulch and plants may slide off, especially when wet or weighed down with snow (it is possible to construct a system of shelves and netting to prevent soil slippage).

The layers of a typical living roof are as follows, starting from the lowest layer which sits on the roof rafters:

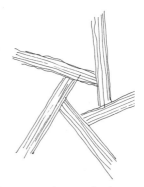

Roof trusses coming together in a vortex.

1. A layer that creates a smooth surface. This can be anything from plywood sheathing or boards to a smooth insulative straw-clay finish on top of rough surface decking such as carrizo (see figure c left) or previous straw-clay layers (figure a). If plywood or boards are used, and the roof has a steeper pitch, it is advisable to create a textured surface to prevent subsequent layers from sliding off (figure b). Nailing on some 2 x 4 boards, or shaping the straw-clay to create horizontal undulations, will help (figures b and c).

2. A waterproof membrane such as bentonite clay with a layer of geotextile membrane to prevent root penetration, or polymer-based modified bitumen, or some other kind of durable, reinforced, and impermeable sheeting.

3. A cushioning layer of corrugated cardboard or carpet scraps placed on top of the waterproof membrane to prevent it from being punctured and to give the vegetation something to take root in.

4. A layer of soil or other organic matter 2 to 8 inches (50 to 200 millimeters) deep, seeded with plants. Rock gardens are of course more appropriate in drier climates.

Roof construction consisting of carrizo decking with straw-clay for insulation used on houses in Xochitl, Sonora, Mexico.

LOW-COST FLAT ROOFS

Roofing is one of the biggest challenges in low-cost construction, since it is usually the most expensive part of the structure. When I worked in Mexico on a housing project organized by the Canelo Project, the roof systems developed were for straw bale houses that cost between 350 and 500 U.S. dollars. The roof types we used

Cardboard box roof construction used on the houses in Aves del Castillo, Sonora, Mexico.

BILL STEEN

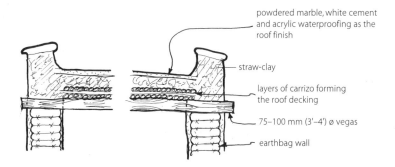

powdered marble, white cement
and acrylic waterproofing as the
roof finish

straw-clay

layers of carrizo forming
the roof decking

75–100 mm (3'–4') ø vegas

earthbag wall

Roof construction using cardboard boxes filled with straw on a concrete-reinforced grid supported by chicken wire.

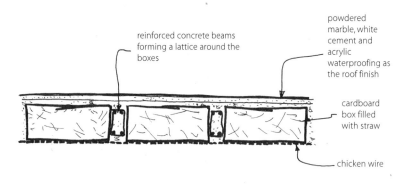

reinforced concrete beams
forming a lattice around the
boxes

powdered
marble, white
cement and
acrylic
waterproofing as
the roof finish

cardboard
box filled
with straw

chicken wire

Roof construction consisting of carrizo decking with straw-clay for insulation and sculpted parapets.

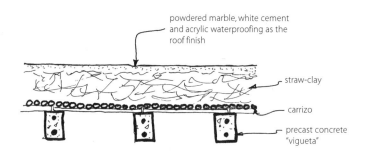

powdered marble, white cement
and acrylic waterproofing as the
roof finish

straw-clay

carrizo

precast concrete
"vigueta"

Precast concrete vigueta roof system with carrizo and straw-clay insulation. Viguetas are short structural supports that span between main beams. Instead of concrete viguetas, short timber poles could be used.

there are also well suited for low-cost earthbag construction.

For the rafters we used 3- to 4-inch- (75 to 100-millimeter) diameter poles, discarded from timber cutting operations because they were too small for convenient milling. The roof surface or decking was constructed of carrizo and covered with two layers of straw-clay mix. The first layer of the mix contained uncut straw to sculpt the parapets and build necessary thickness (6 to 7 inches) to provide reasonable insulation value. The second layer contained finely chopped straw and clay to even out the surface and prevent puddles. The roof finish was a mix of powdered marble, white cement, and an acrylic waterproofing compound as a final coating (see the discussion of waterproof finishes in chapter 6).

Another low-cost alternative option for a dry climate (used by the Canelo Project in Mexico) is a roof made out of cardboard boxes filled with straw or other insulating material, which are laid flat on some form of inexpensive joist or rafter arrangement, or chickenwire for support, as shown at left.

ROOF INSULATION

Many materials can be used to insulate a roof, the lower-cost options being straw-clay (a thick clay slurry mixed with a lot of straw, as shown in figures a and c on page 70) or straw bales, as described below. A slightly more expensive option would be insulation blown in between rafters, which could be recycled cellulose, hemp, wool, or coconut fiber. As for insulating a flat roof, if pumice is locally available it can be used by placing 8 to 12 inches (200 to 300 millimeters) or more of small pumice (only small particles of pumice wick moisture) with 6 inches of earth on top to allow for planting. An insulated roof has to be framed in at the edges like a box to contain the large volume of materials.

Another type of insulated living flat roof uses straw bales as insulation. The slope should not exceed 30 degrees. As in the other living roofs discussed, a water-proof membrane is placed between the decking and the straw. Then straw bales are laid flat, leaving a gap of 3 to 4 feet (1 meter) at the edges where flakes of straw can be used to taper the roof to the height of its frame. When the bales have been laid, cut all the strings to loosen the straw. For the first winter and spring, leave this exposed to allow the straw to soak up the moisture, then cover it with a very thin layer of aged compost and sow flower seeds. The best plants for such roofs have a shallow root network and retain moisture well, for instance strawberries. Imagine having a roof full of strawberries!

Natural Finish for Flat Roofs

A durable finish for flat straw-clay roof surfaces can be made with a capping of two coats of ½-inch (12 millimeter) lime render (1 part lime to 3 parts sand).

Here are two recipes for waterproof finishes to be painted on the surface:

1. Dissolve 2½ ounces (70 grams) of alum (aluminum potassium sulphate) and 2½ ounces (70 grams) of salt in 2 cups (½ liter) of water. Add this mixture to 5⅓ quarts (5 liters) of water and mix in 1/16 sack of lime. Use this to paint over the finished lime-plastered roof surface.

2. Apply five coats of dissolved alum and soap, alternating these in the following way:

   Day 1: Dissolve 14 ounces (400 grams) of soap in 4 cups (1 liter) of hot water and brush on roof surface.

   Day 2: Dissolve 14 ounces (400 grams) of alum in 4 cups (1 liter) of hot water and brush on roof surface.

   Alternate these for five days, and reapply every year or two.

Recipes make enough finish for approximately 30 square meters.

6

WEATHERPROOFING AND FINISHES

As with any building, keeping the water out of an earthen house is one of the greatest concerns of builders. As the cob builder's proverb of Devon, England, says: "Good shoes, good hat, and a coat that breathes." This is what an earthen house needs to survive for many decades.

Good shoes, to raise the building sufficiently off the ground—a sturdy, well-drained foundation.

Good hat, a generous overhang to protect the walls from erosion from the rain.

A coat that breathes, a plaster that allows the passage of moisture.

This chapter applies to internal and external earthen plasters (often called "renders," when exterior) for earthbag domes and other types of earthbag houses, as well as compatible finishes for benches, ovens, stoves, or any other earthen structures.

In the earthbag construction system, the wall surface is never the bare earth, but whatever material the bags are made out of. Rendering an earthbag house is necessary for several reasons.

If the bags used for the construction are polypropylene, they need to be covered within the first two months of exposure to ultraviolet light (direct sunlight), as UV light makes the bags deteriorate, exposing the material inside the bag. If the material inside the bag has 10 percent or higher clay content and the structure was properly tamped or rammed throughout the construction process, the walls should remain solid and stable even when the bags have deteriorated. However, if the fill is of a loose composition, such as silt, sand, gravel, or pumice, the bags must be covered. The type of covering used will depend mainly on the climate and on the design of the house. For example, a dome in a rainy climate will require a plaster that is water resistant, such as lime, papercrete, or cement-stabilized soil. In extremely wet climates a waterproofing layer on the top part of a dome or vault is essential (see the chart on page 77). If the water-resistant render becomes saturated with water and has no form of sealant or impermeable membrane, the moisture will go down with gravity through the earthbags. If the interior is covered with an earthen plaster, this will quickly absorb water and come apart. On the other hand, if the house is designed

Facing page: The Hart's dome covered with papercrete render.

with a conventional roof and wide over-hang, and is raised up from the ground by the foundation and stem wall. The walls do not need a water-resistant render, being protected from the driving rain, and an earthern render can be used externally.

If the bags used are made out of burlap (hessian), they will also need to be covered, but they have a longer exposure life than polypropylene.

Another reason for rendering the exterior of a wall constructed out of earthbags is that the surface has many grooves and seams that in some extreme weather conditions can be penetrated by rain, which must be prevented of course. The grooves can be extremely useful when applying a render, however, providing a "key-in" area for both external and internal plasters. If for aesthetic reasons the ribbed pattern on the wall is desired, a render can be sprayed on, retaining the pattern of the bags.

Earthen Plasters

If any plaster is used on a bare earthen wall that contains clay, the plaster should be a coating that breathes, allowing any moisture that enters the wall to escape. Earthen plasters have been used extensively in many countries for many centuries. As well as being used as a finish coat for adobe, cob, or straw bale they also make an excellent covering for earthbag walls that have a roof overhang. With earthen plasters, whatever moisture does penetrate into the walls will be absorbed automatically by the clay, due to its hygroscopic

(water-thirsty) properties, and then released to the outside. There may be no other building material capable of regulating moisture levels as effectively as clay, which continually absorbs and releases moisture in response to the humidity of the home. With thick and solid rammed earth, adobe, or cob, an external render may not be necessary.

In Devon, traditional cob houses have survived for centuries without any plaster coating (they say it takes one hundred years to wear away one inch of cob). Provided that cob walls are protected from actual erosion caused by the abrasion of driving rain, there is no necessity for external render, because moisture will evaporate very quickly from an exposed cob surface. With earthbag construction, however, rendering the external surface is a necessity, as emphasized above.

Among the advantages of earthen finishes are these attributes:

- moisture-control
- fire-resistance
- odor-absorbent
- nontoxic
- when dry, are unaffected by frost
- aesthetically pleasing

Earthen walls covered with earthen plaster may give the impression of having grown out of the landscape. Their subtle colors, complementing those of the ground that surrounds them, can add greatly to the charm of the countryside.

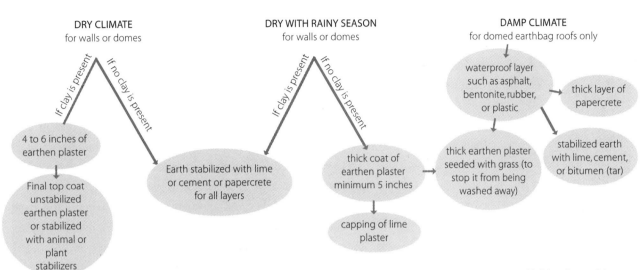

DRY CLIMATE
for walls or domes

If clay is present

If no clay is present

4 to 6 inches of earthen plaster

Final top coat unstabilized earthen plaster or stabilized with animal or plant stabilizers

Earth stabilized with lime or cement or papercrete for all layers

DRY WITH RAINY SEASON
for walls or domes

If clay is present

If no clay is present

thick coat of earthen plaster minimum 5 inches

capping of lime plaster

DAMP CLIMATE
for domed earthbag roofs only

waterproof layer such as asphalt, bentonite, rubber, or plastic

thick layer of papercrete

thick earthen plaster seeded with grass (to stop it from being washed away)

stabilized earth with lime, cement, or bitumen (tar)

Finishes for earthbag domes and walls.

Meanwhile, among the disadvantages of earthen plasters are that they have a low structural resilience, therefore the design of the house is critical. If they are made with earth that possesses a high clay content, or with very fine sand or silt as the filler and little or no straw, the plaster may be easily eroded. Also, earthen plasters are affected by frost in cold climates; if moisture is allowed to penetrate the surface, it will expand and contract as it freezes and thaws, breaking up the plaster. In cold areas it is a good idea to cap the earthen plaster with a lime plaster.

In locations where the plaster finish is steadily eroded by weathering, it has to be maintained on an annual basis. This can be turned into a fun ritual.

Application

Earthen plasters are incredibly flexible to work with, allowing everyone to find a personal way of mixing and plastering. They can be applied in two or more stages. The purpose of the first layer is to fill in large gaps and crevices and to build up the main bulk, creating a fairly even surface for the smoother plaster to go on. This first layer contains straw that is either uncut; direct from the bale, which provides an interwoven stability; or finely chopped, which is easier to mix in larger quantities and is also great for building up thickness. Long straw in the mix creates a plaster-reinforcing network and helps to fill out large holes or build up bulk where it is needed, as in sculpting the sills around windows.

An earthbag wall contains indentations between the courses of tamped bags, making it easier for the plaster to "key into," or adhere. No plaster-reinforcing lath is necessary, as this can interfere with the earthen plaster being keyed into the wall. A good plaster mix is already well reinforced by straw. As this first coat of plaster

For earthbag houses with conventional roofs providing appropriate overhangs, no waterproof layer is needed.

is applied, it should not be smoothed out too much, but instead left rough, or into finger-sized holes so that the next layer of earthen or lime plaster will adhere. If the plaster does not stick to the wall, pound wooden pegs into the grooves between courses; these can also be inserted during construction.

The second coat of earthen plaster is more refined and can be very thin, just enough to allow a final smoothing out. This can be a mixture of finely sieved sand and clay and, if desired, finely chopped straw, along with wheat flour paste or another stabilizer.

Stabilization and Alternatives

Stabilizers are generally used to make renders or plasters more durable and resistant to moisture. They are the "glue" that can be used to bind filler particles such as sand, earth, gravel, or fibers such as straw. Stabilizers can be used as additives to earthen plasters if the earth mix does not contain enough clay to provide the binding force and moisture resistance required.

Stabilizing earth is a very complex process, since not all stabilizers are effective with all soil types, and there are many factors that influence compatibility, including clay content, soil particle size and type, pH balance, and climate. Other factors that must be taken into consideration are the type of application and the reliability of periodic maintenance, as well as aesthetics and cost.

As explained below, if an earthen plaster is well mixed and contains a good distribution of clay, aggregate, and straw, adding stabilizers can be largely avoided through design features. But earthen walls and plasters must be stabilized if the earthen mix does not produce enough binding strength (that is, does not contain enough clay). Yet remember, houses built out of earth need to move slightly over time, and they are especially sensitive to dampness and temperature, so they need to breathe, releasing moisture. Modern stabilizers and sealants result in severe damage to earthen buildings, because they tend to restrict movement and permeability to moisture. Lime-based or other natural stabilizers do allow the walls to breathe and move, adhere much better to earthen walls (not requiring chicken wire or stucco netting), and produce more porous finishes than Portland cement, the most widely used industrial stabilizer, which is an expensive and environmentally controversial material, as discussed below.

It is possible to improve the characteristics of many types of soil (especially sandy soils) by adding stabilizers. These stabilizers can be used in the earthwalls themselves or in their skin as a surface protection. Due to the vast variety of soil types, stabilization is not an exact science, and research is continuous. According to CRA-Terre's *Earth Construction,* "The best known and the most practical stabilization methods are increasing the density of the soil by compaction, reinforcing the soil with fibers, or adding lime, bitumen, or cement." (Houben & Guilland 1994) When choosing the best stabilizer for a particular soil, many factors such as the clay content, acidity, and texture must be taken into account, and many samples must be made prior to construction. Again from CRATerre, "It is particularly unfortunate that many practitioners of systematic stabilization do not know, or do not appreciate the original characteristics of a soil, and start about stabilizing soil with undue haste, when it is not particularly useful." The wrong stabilization may do more harm than good.

As a case in point, cement can be a real enemy of earth architecture, apart from the few selected applications such as bond beams on top of walls, compression rings in domes, and stabilization of sandy soil when making soil cement, which can be applied as a render to earthbag domes in areas with high rainfall. But cement should never be applied on top of an earthen plaster, as this will eventually crack and peel off. As a general rule in earth architecture, never place a hard, modern, nonbreathing material on top of a more flexible surface, as this will never form a solid bond and will eventually separate, as well as create condensation and other moisture problems.

Buildings constructed of local stone, earth, and lime cause far less environmental damage than concrete and steel. Earth and stone are reusable, and old, dry lime render is chemically limestone again, just as when it was first quarried and can also be reused (see the diagram on page 81 showing the lime life cycle).

In industrialized societies, an increasing number of people have been affected by "sick building syndrome," which is a form of poisoning harming people who live and work in buildings with insufficient ventilation in which toxic vapors are given off by artificial, chemical-intensive building materials and paint. Cementitious finishes release carbon dioxide, a greenhouse gas, as they cure. Over the longer term, they seal in moisture and therefore can cause air-quality problems and propagation of mold.

Cement is also a more expensive material, because it contains higher levels of embodied energy, although unfortunately in some countries where demand for lime is very low (as in the United States), the healthier option can be as expensive as cement. The benefit of lime is its superior quality, and when properly applied it is worth every additional effort and expense.

Sources of stabilization include:

- vegetable stabilizers
- processed natural binders
- animal stabilizers
- mineral stabilizers

If a house is built in rainy or damp climates, the walls could be prone to severe weathering from driving rain and frost. With effective planning and the right detailing, originating all the way back in the initial design and building process, it may be possible to forgo the use of stabilizers and instead use a plaster made of just earth.

There are several ways to protect earthen walls without using lime or Portland cement for stabilization. During the design and detailing process, familiarize yourself with the direction of the sun and driving rain, and plan accordingly to provide a roof with a large overhang (a minimum of 18 inches or 450 millimeters on the least problematic side). Extra-large porches can be integrated into the house layout on the sides that are most vulnerable from wind-driven rain. The more protected the walls are, the less protection that plaster needs to provide. Remember, each wall can be slightly different; the most exposed wall can be the only one capped with lime plaster. Always slope surfaces away from the ground around the base of the house as well as sculpted windowsills, alcoves, or seating to shed the water. Design these to extend out at least 2 inches (50 mm) from the wall, with a drip edge (see the diagrams on page 61). Temporary screens could be installed as necessary to stop the seasonal rain, or to prevent melting snow from eroding the walls, for example a bamboo or willow screen. Or if bad weather is year-round, on the side where the most snow settles or the strongest winds blow, a high stone wall could be incorporated into the base of the walls. Permanent screens could also be built to keep driving rain off the wall, keeping a distance of a couple of inches away from the wall to allow for ventilation.

Adobe walls are traditionally protected by earthen plasters that are annually "topped up," but allow the wall to breathe. This can also be applied to earthbag as well as straw bale houses. Other ways of minimizing the penetration of moisture include sealing the earthen plaster with oil or whitewash (see list of sealants starting on page 92); stabilizing the earthen plaster with lime, wheat-flour paste, or other plant, animal, or mineral stabilizers (see the list starting on page 81); or putting a final cap of lime—a more durable, but still breathing surface, over the earthen plaster.

Prior to any construction, it is crucial to know your soil. See chapter 7 for a discussion of soil-testing methods.

Stabilization is not compulsory. If the soil contains enough clay, you can ignore

Caution!

Stabilizers function as binders in the mixture but are not sufficient on their own. To make them solid and hard, they need to be combined with other fillers, for example nonexpansive particles of soil such as sand, silt, gravel, or fibers.

Never layer a rigid, modern, non-breathing material on top of a softer, breathing surface. The layers will eventually separate.

the whole question quite satisfactorily. As the CRATerre publication *Earth Construction* explains,

> there is clearly a tendency at present to the over systematic use of stabilization, which is regarded as a universal panacea for all problems. This attitude is unfortunate, as stabilization can involve considerable extra costs, ranging from 30 to 50 percent of the final cost of the material. Furthermore, stabilization complicates the production of the material. It is thus advisable to insist that stabilization is only used when absolutely essential and that it should be avoided where economic resources are limited.

Vegetable Stabilizers

The following vegetable-based materials will serve as effective stabilizers for earthen plasters:

- oils—coconut, linseed, and cotton, which need to be in "boiled" form to speed up the drying process
- juice of banana leaves, precipitated with lime, improves erosion resistance and slows water absorption
- prickly pear juice (found in the southwestern United States)
- wheat flour paste, or any starchy material

Linseed oil can either be brushed on the finished surface of an earth plaster, as is often done with the final layers of earthen floors (see chapter 7) or can be mixed into the final batch of plaster itself.

To make prickly pear juice, the cactus has to be boiled until very soft, then its juices squeezed out. The resulting soupy liquid is then combined with the clay and soil mix that is to be used for the plaster finish. Like any other stabilizer, this has to be tested prior to use, as it reacts differently with different soils.

Wheat flour paste is an inexpensive stabilizer for earthen plasters that is used by natural builders throughout the United States.

Harvesting prickly pear cactus.

Prickly pear cactus being boiled.

I learned it from Carol Crews of Gourmet Adobe, while we were making clay paints, and it can be applied like a paint to earthen plasters or floors. It can be made using commonly available flour, as described on page 96.

To make a plaster without using lime or clay as the binder, you can combine sand with manure and wheat flour in the following proportions:

4 parts flour paste (see page 96)
3 parts sand
2 parts manure

Other processed natural binders that can be used to stabilize earthen plasters when too little clay is present include wallaba resin; rosin from oily pine resins, obtained during distillation of turpentine; copal, made from tropical tree resins, added in a proportion of 3 to 8 percent for sandy soils; gum arabic, from the acacia tree; and molasses.

Animal Products as Stabilizers

Among the most popular animal stabilizers in use today are the following: eggs, blood, and casein (dried milk) as proteins; urine and manure, as uric acid; casein (dried milk); and glue made from animal parts or byproducts. Termites secrete a chemically active substance, and termite hills stand up well to rain. Their soil can be mixed with other soils for the production of blocks that adhere effectively; perhaps this soil would also stabilize earthen plasters.

To make casein glue to stabilize earthen plasters, soak about 1 ounce (25 grams) of casein powder and ¼ ounce (8 grams) of borax in enough water to form a putty (to make casein, see page 98). The putty can then be diluted with water to a consistency suitable for mixing with the soil ingredients of the plaster. For more casein recipes see the end of this chapter.

Mineral Stabilizers: Lime

For thousands of years in Europe, lime has been used as a mortar for stone or brick construction; as an exterior or interior plaster, when mixed with sand; and when mixed with water, as a white paint commonly known as whitewash. In the mid–nineteenth century, cement and gypsum unfortunately became more common building materials. Lime was slower to build with, and required artisinal skills and good climatic conditions during application, but produced durable and attractive results. Lime plasters and finishes harmonized with seasonal changes in humidity and temperature, like clay preventing an overly dry or wet atmosphere by evaporating away excess moisture or absorbing it as necessary, fostering a healthy living environment. Simultaneously, due to its alkalinity, lime does not allow the growth of mold on the walls, therefore creating healthier conditions in wet climates.

Using lime as the binder in renders, plasters, and mortars works best when sand is the filler. Lime is not the most effective stabilizer to use with an earthen plas-

An earthen house with partial lime render, Peru.

samples before beginning to plaster an entire wall.

LIME PLASTERS

Historically most buildings in the United Kingdom used lime for interior plasters and exterior renders. Cob buildings, if rendered at all, were traditionally covered with a lime-based render applied directly to the cob. This render consisted of lime, which serves as the binder, and sands and/or aggregates, which provide bulk at low cost, and which control shrinkage. It is best to use aggregate that has a good range of particle sizes.

Fiber—for instance, hair—was often added to the traditional mix to minimize the shrinkage cracks that often occurred. Cow hair was preferred, but was harder to obtain, so goat hair was frequently used. Some straw bale houses that are being rendered in lime use straw as the fibers.

Using lime plasters has several advantages. When lime used in buildings has set, it turns back to limestone, which is chemically the same as the lime that is quarried,

ter or adobe mix that contains clay, because the lime and clay seem to "compete" as binders. Ideally the mixture should turn into a creamy paste, which will result when it is dry in a finish that is more wear- and water-resistant than clay alone, but in reality this is rarely achieved due to variations in the pH of clays. Therefore combining clay and lime can make a crumbly mix, and when used for stabilizing a wall material such as adobe, can reduce its compressive strength.

The theory is that, since lime is alkaline, it combines best with acidic soils. The higher the pH of the soil, the more lime is needed to stabilize it. Lime apparently does not react well with alkaline soils, therefore carrying out tests is crucial to finding out the soil's behavior. If the mix of earthen plaster and lime does not turn smooth and creamy, the addition of either a more acidic ingredient such as manure or organic soil may help. But care needs to be taken; always test the mixes by making

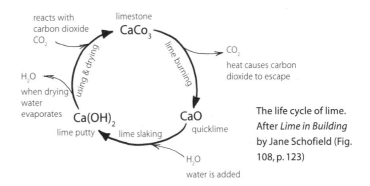

The life cycle of lime. After *Lime in Building* by Jane Schofield (Fig. 108, p. 123)

Slaking lime in Xochitl, Mexico.

since the water has been driven off, it becomes very "thirsty" and reacts dramatically with water (even water in the air, or in the skin or eyes).

When the quicklime is soaked in water, it turns to slaked lime, due to hydration. This process produces a lot of heat, as the mixture boils violently. This is known as slaking. The resulting slurry is calcium hydroxide ($Ca(OH)_2$), known as lime putty.

When applied on a wall surface and therefore exposed to air, the lime reacts with the carbon dioxide in the air to form limestone again, that is, calcium carbonate, ($CaCO_3$). This happens as it dries on the building. The water evaporates, and the lime hardens through carbonization, thus completing the cycle (Schofield 1994).

Making Lime Putty (Slaking)

For this task it is necessary to wear protective clothing, including gloves, goggles, and mask. Since quicklime is caustic, it can burn your skin, and during the slaking process the mixture can spit violently while boiling.

It is recommended to use 2 parts water to 1 part quicklime.

Slaking has to be done outdoors with a metal container, making sure it is not placed on any flammable material, because the container will get extremely hot. The water is poured in, then quicklime is added slowly to the water, with one person stirring at all times (a backhoe is the best

Caution!

Always add the quicklime to water, *never* add water to quicklime, as this can cause an explosion!

therefore it can be reused if the building is destroyed and does not involve any processing that will harm the environment.

In terms of healthy-home considerations, lime plasters and washes usually prevent condensation and an overly dry atmosphere, because the moisture is absorbed and then given out, contributing to higher air quality. Also, due to its alkaline properties, lime prevents mold problems.

Lime is ideal for use both as an exterior render or as an interior plaster or wash.

The chemical process of obtaining lime involves heating calcium carbonate ($CaCO_3$) from limestone or shells to 1,200° Celsius. This can be done in kilns (shells can also be heated in a pile covered with cowpats and coconut husks). The heating causes carbon dioxide (CO_2) and steam (H_2O) to escape, and quicklime, or calcium oxide (CaO), remains. At this stage,

tool for large quantities). As more quicklime is added, the mixture starts to boil and bubble. When that happens, stop adding quicklime and keep stirring until the mixture ceases bubbling, or else the lime will burn and be of lower quality. The better the quicklime mix, the faster the hydration process occurs. The mixture should be stirred continuously until all the lumps are broken down, and until the mixture has cooled down and is of a creamy consistency. It can then be sieved through a $^1/_{16}$-inch (2 mm) sieve to take out the unburned limestone pieces.

The resulting mixture is lime putty. When left standing, the lime sinks and the water sits on top. For best results it is recommended to leave it to mature for at least three months prior to any use to ensure that all the calcium oxide has hydrated, and the longer it sits the better it will be, though in Mexico we used it after only one month. While curing it must remain in a sealed container to prevent it from drying out, which is the carbonization process whereby the lime putty will turn back to limestone, losing its binding properties and therefore becoming useless as a plaster.

Lime putty made from quicklime is by far a superior material to the bagged hydrated lime available from garden supply stores, which when soaked in water produces a putty that is generally not as good. The problem with the bagged material is that it might have been sitting in the shop

for a long time, which means many of the particles in the mix will have already reacted with moisture in the air and carbon dioxide to form limestone, making them inactive and therefore weakening the mix.

To make lime putty out of bagged hydrated lime, mix the lime with water in a bucket to form a paste, cover this mixture with more water, and put an airtight lid on the storage container. Store for several weeks before using. Make sure it does not dry out by adding water to it occasionally. The longer it sits the better it gets. If this bagged lime is not very fresh or of the best quality, mix 1 part of the resulting putty with 2½ parts sand (instead of the 3 parts suggested below) to compensate for the inactive particles. To further strengthen bagged hydrated lime, add some of the more expensive "hydraulic" lime, which sets faster and is more frost resistant after

Earth houses with lime render on a street in Cuzco, Peru.

just three days instead of two weeks. A good compromise is ⅓ hydraulic lime to ⅔ hydrated lime.

Making Lime Plaster or Render

In order for a lime plaster to dry and become limestone again, it needs to give off all its moisture and draw in carbon dioxide; therefore, it cannot be applied in thick layers. When applying a first coat of lime to an earthbag surface, much crushed aggregate or straw needs to be added to allow for drying in the deepest areas keyed in between the bags. It is therefore better to first cover an earthbag wall or dome with earthen plaster to fill in those deepest voids before applying subsequent coats.

Lime plasters can vary enormously, depending on the type of lime, aggregate, and particular use. The proportions of lime to sand may vary between 1:2 for a smooth, fine finish to 1:5 for a rough first coat. For greater strength in a plaster, the sand particles used should be well graded, ranging from very fine to coarse in one mix.

As mentioned earlier, the only difference between the mix for interiors and exteriors is the size of the sand or other aggregate. A finer mix is best for the inside, and a coarser for the outside—from very fine dust to as large as 3/16 inch (5 millimeters) for interior plaster, and up to twice as large for exterior renders, which should be angular in texture to reduce the penetration of moisture. Limestone aggregate is particularly recommended, because then the filler binds especially well. If sea-dredged sands are used, they require washing several times in clean water to remove salts.

The most popular proportions of lime putty to sand is:

1 part lime putty to 3 parts sand

Caution!

- The durability of lime depends upon the quality of lime and the right mix, as well as the quality of the application and the drying conditions. Variables in the mix of lime and sand—including proportions and the particle sizes—are crucial, as are the weather conditions during the application process.
- Apply in thin layers and make sure the plaster is well keyed into the layer below.
- If using lime to stabilize soil, always test your mixture in advance. If not enough lime is added, the compressive strength can be lower than that of unstabilized earth.
- Do not overwork lime plaster with a metal trowel. This makes lime come to the surface and can form a hard crust over a softer backing, weakening the plaster.
- Before application, make sure you cover all metal surfaces, as the lime can stain.
- Do not put lime plaster on top of gypsum, wood, or latex.

When preparing the plaster, it is important to mix and beat the lime putty for a long time on a wooden or plywood surface with wooden mallets and posts to get it to become more plastic, and then work it well into the sand. The more you mix, the better the plaster.

Application

Make sure there is good adhesion between the lime and the earthen plaster underneath. (Remember, before the earthen plaster dries, to scratch its surface and make finger-sized holes for "keying.") Traditionally lime-sand plasters have been applied in two coats of not more than 10 millimeters thickness each. When applying lime plaster, the earthen plaster (or the earth wall) underneath must be fully dried out then slightly dampened to help the lime grip the earthen surface. It is best to use limewater for this (made by dissolving 2 to 3 percent of lime in the water).

After application of the first coat it should be scored to provide key-in areas for the second coat. This can also regulate the shrinkage cracks. Noticeable shrinkage in the first coat can serve as a warning that either the lime is too fresh, the mix is too wet, or the plaster is applied too thick; the second coat should be thinner, as a safeguard.

The second coat should be applied when the first is "green-hard"—that is, too hard to dent when pressed with a knuckle, but soft enough to mark with a thumbnail.

Knock off any lumps around the score marks and spray the first coat with limewater before re-coating. When the second coat is hardening, it may be worked over again to improve the finish, remove any rough spots, and push closed any small cracks. More detailed instructions can be found in the book *Lime in Building: A Practical Guide* by Jane Schofield (see the bibliography).

Protect the plastered surface from sun, wind, and frost to prevent it from cracking during the drying process. It needs to dry slowly. Ideal conditions are humid, cloudy days with virtually no wind and a slight drizzle. One way to provide protection from the sun is to hang wet cloths a few centimeters away from the plastered surface. Frost can also be very damaging during the drying period, and frost protection may be needed for a minimum of two weeks after the lime has been applied.

Pozzolanic Additives to Lime Plaster

Pozzolanic material can be added, ground up into a powder, to speed the setting of lime and to help the lime mix set deep inside the wall, which is necessary when using lime as a mortar in stone or brick construction, or as a render directly on top of earthbag walls where there are deep indentations. Some examples of pozzolanic materials are crushed clay bricks, clay tiles, shales, potash, and pumice. Pumice is a naturally occurring pozzolan from volcanic areas. These pozzolanic materials

Recipes for Lime Mortars, Renders, and Plasters

—from the Earth Building Association, Devon, England

Mortar for Bedding

(can be applied as a first coat to earthbag walls)

12 parts coarse sand

3 parts lime putty

gauged with 1 part pozzolanic additive (for example, brick dust)

Roughcast Render: Backing and Finish Coats

2 parts coarse sand

1 part grit (up to 4 millimeter diameter)

1 part lime putty

1 bucket mix to ½ bucket of teased hair. Omit hair in the final thrown coat. Hair must be teased out with carding combs to remove the large clumps.

Smooth Render

1 part coarse sand

1 part grit (up to 4 millimeter diameter)

1 part fine sand

1 part lime putty

1 bucket of mix to ½ bucket of teased hair

Lime/Manure Render

1 part lime

4 parts wet cow dung (a few days old)

1 part sandy earth

Lime/Quark Render

4 parts lime

1 part fat-free quark (to make quark, see the recipe on page 99)

10 parts sandy earth

Tallow, an animal fat, increases water resistance and adhesion. Ten percent by weight of melted tallow can be added to lime; this can also be replaced by linseed oil.

speed up the setting of lime due to the reactive silica present in them, which combine with lime at ordinary temperatures in the presence of water to form stable, insoluble compounds with cementing properties. The rate of reaction is increased by increasing the fineness of the pozzolanic material (Spence & Cook 1983).

STABILIZATION FOR WATERPROOFING

In addition to the water-resistant layers on domes, extra protection is needed in extremely wet climates. Imperviousness will help to reduce water erosion, swelling, and shrinking when the plaster material is subject to successive wetting and drying cycles. For waterproofing an earthen plaster, a material that is unaffected by the water that fills the voids, pores, and cracks is required.

Bentonite clay is a material that is dispersed in the soil and that expands upon the slightest contact with water and prevents the infiltration of pores. This has recently been tried as a waterproofing layer in living roof construction (see chapter 5), but its use is still at an experimental stage. When used as a waterproofing layer, it needs to be weighed down with a large amount of soil, as it expands and moves with absorption of moisture.

Bitumen is a mixture of hydrocarbons and other materials, either occurring naturally or obtained by distillation of coal or petroleum. For instance, tar and asphalt are bituminous. The use of bitumen as a stabilizer is very ancient, dating back at least to Babylon in the fifth century B.C.E., where it was used for making mortar or laying unbaked molded bricks. Bitumen mixed with soil acts as a water-repellent, reducing penetration and surface erosion from wetting, therefore serving more as a waterproofing element than a binder. It is most successfully used with granular soil, in which it improves durability, but is also widely used with clay soils, for instance in the manufacture of adobe bricks. Stabilization with bitumen is in fact most effective in a process involving compression, as in production of compressed clay blocks. To stabilize adobe, 2 to 3 percent of bitumen can be sufficient, and sometimes as high as 8 percent is required. If used as a stabilizer, bitumen either must be mixed with solvents or dispersed in water as an emulsion. If used with soils containing high proportions of clay, a larger amount is required due to greater resistance to mixing. To obtain an even distribution, large quantities of water need to be used. Solvents that can be used include diesel oil, kerosene, naphtha, and paraffin (a mixture of 4 to 5 parts bitumen to 1 part paraffin oil).

Note that these solvents cannot be used in the rain and are flammable. Also, bitumen stabilization is not effective in acid, organic, or salty soils.

A simpler, more natural weatherproofing solution can also be made by mixing clay with pure linseed oil and applying it onto earthen plasters.

Stabilization with Cement

Cement should only be used as a stabilizer in plasters as a last resort in cases where the soil does not contain enough clay and other, more natural stabilizers aren't available. For instance, if lime is not available, soil stabilized with cement can be used on polypropylene earthbags containing sandy fill with little or no clay. Cement interferes with the binding forces of clay; therefore, care needs to be taken when deciding on the quantity of cement to be added to your soil; the higher the clay content, the more cement is needed. Somewhere between 3 and 10 percent will probably be appropriate, but tests should be carried out to determine the necessary quantity more precisely (see chapter 7 for more information).

If cement is used, commercially available Portland cement has the least embodied energy (that is, it requires the least energy for processing and preparation). It is made of burned lime and highly reactive silica. It introduces a three-dimensional matrix into the soil and results in a filling of the voids with an insoluble binder, which coats the grains and holds them in an inert mass.

While a cement-stabilized render can in some cases be used over earthbags, for instance soil cement or conventional stucco finishes, *never* use cement plaster on top of adobe or cob walls or a nonstabilized earthen plaster!

As emphasized repeatedly in this chapter, when cement plaster is applied on top of earth, it forms a brittle, rigid surface that is impervious to moisture. The effectiveness of cement-based plaster is dependent upon the rigidity of the wall beneath it, since the cementitious finish itself forms a rigid, relatively brittle shell. An earthen wall continues to move over time; this is normal, and is even beneficial in seismically active areas. Due to the different properties of cement and earth, temperature changes and moisture cycling tend to produce cracking in the cement render. These hairline cracks could be almost invisible, but once the waterproof finish is compromised, any moisture drawn in through these tiny cracks will be trapped, unable to evaporate, and will start wearing away at the softer material behind. Given enough time, big cavities can be worn away, even across the whole width of a wall. This would not be such a problem if the damage were detected early enough, as the wear is very gradual. But the surface of the cement-based plaster does not wear very noticeably, therefore hiding the problem until entire chunks of wall cave in or collapse.

Such moisture damage is most likely to occur where hard, modern materials are applied as a finish to a building whose underlying structure is made of softer, more flexible earthen materials. A lime-based render or earthen plaster acts more like blotting paper, absorbing and releasing moisture relatively freely. In addition according to the *Devon Earth Builders,* small cracks may be closed by redeposition of soluble material from the lime or clay.

Carol Crews of New Mexico's Gourmet Adobe explains the large lesson learned at the famous St. Francis de Assisi Church in Ranchos de Taos:

> When it was coated with cement stucco in 1967 this plaster cracked and allowed the moisture to penetrate deeply into the adobes, but the relatively impermeable stucco prevented the adobes from drying out again. Large sections of the buttress had to be rebuilt, so the community has now gone back to the annual renewal of the mud plaster, which not only keeps the church building in beautiful condition, but strengthens neighborhood ties as well. (Kennedy 1999, 95)

It is surprising that even after experiences of this kind, the U.S. Uniform Building Code continues to stipulate the use of cement plasters on top of the adobe walls in several of the Pueblo Indian villages in New Mexico. With cement-based renders, a great deal of work is still required for maintenance, but often the result is a patchy surface, since repairs with cement plasters are usually visible. Instead of being

Chunks of cement plaster cracking off an adobe wall of the monastery in Abiquiu, Santa Fe, New Mexico.

allowed to simply give each house a thin coat of earthen plaster or whitewash once a year, the buildings have to be carefully and expensively repaired.

As a general rule: *Renders should have a permeability equal to or higher than that of the wall material.*

Application of Stabilized Renders

With the addition of some stabilizers the render applied to the domed or vaulted parts of the structure becomes more brittle, and cracks can occur through expansion and contraction with extreme temperature changes, as discussed above. This movement can be controlled through the fragmentation of the render mass. By placing the render in small "patties," a textured finish can provide thermal variation throughout the whole surface of the structure, creating air movement due to the temperature differential between the sun zone and shade zone within the render itself, never allowing the surface to overheat. As one side of a rounded patty heats up, the other cools down. This surface has been used for centuries in African villages and is prevalent in nature, for example in the scales of a fish or the trunks of trees.

Application

Once you have chosen the right stabilized mix (clay, lime, cement, etc.), apply a "scratch" coat to fill in large cavities and create the desired overall shape. Leave any

Top: The textured, stabilized render applied to the waterproof layer (roofing felt) on a vault in California.

Inset: Textured surface of a vault.

Bottom: Fragmented "patties" of cement-stabilized soil placed directly on the surface of the earthbag dome, California.

irregularities in the surface so the second coat has somewhere to key into. To achieve the bubbly effect, patties of stabilized soil are placed like roof tiles, overlapping each other, starting at the base (like laying tiles) and working up the structure. Stagger the cracks so water will run down the grooves (see the photo on page 91).

Interior Finishes

There are many ways of finishing the insides of earthbag walls, but whichever method is used for the final coating, it is best to coat the uneven earthbag surface with an earthen plaster to fill in any large cavities before the other layers are applied.

As discussed above, lime mixed with very fine sand is a great material to use on top of interior earthen plasters. Another is gypsum. Gypsum is a naturally occurring soft rock or powder. It is converted to "plaster of Paris" by heat. Gypsum is readily available at building supply stores, and it can be applied directly on top of earthen plaster, since it is breathable. Due to its softness, it can only be used on interior walls. It doesn't shrink or crack when dry and sets very fast, which can be an advantage or disadvantage—you need to work fast when applying it to a wall, but the technique is not hard for nonprofessionals to learn. Gypsum is acidic, and has a low embodied energy compared to Portland cement, but in premixed form it is relatively expensive. Gypsum plaster can also be mixed as 1 part gypsum to 2 parts sand for more texture and lower cost, or it

can be combined with perlite for a more lightweight, better-insulating mix. Pigment can be added for a more colorful outcome. Gypsum can also be mixed with lime putty to create a faster-setting plaster than lime alone.

As an alternative to two coats of conventional plaster for walls and ceilings, clay mixed with sand and fiber may be more appealing. (See chapter 7 for more on mixing clay plasters.) Clays come in many different colors, and beautifully colored finishes can be achieved using only the earthy clay colors, for there is seldom a problem with the colors clashing. Mica can also be added to a final coat of clay paint or plaster, which will add a glittering texture.

Earthen plasters are usually applied with hands or a wooden trowel, but for larger projects you can use a hand-held spray gun powered by a gas-driven compressor. Screen the mix through a ⅛-inch screen to eliminate any lumps that might clog the machine. This spray mix is often made with wheat flour paste as a stabilizer (see page 96 for the wheat flour paste recipe).

Sealants

Sealants can also be used as nonstructural stabilizers, but they are called sealants because they seal the earthen plaster—that is, they are not mixed into the plaster during the construction process but are applied like a paint or a finish plaster once the original plaster has dried.

There are two types of sealants: those that form a skin or a shell, and those that penetrate deeply into the earth. Sealants that form a skin or a shell are fine, as long as they breathe (lime is an example of one that does). The main problem with "skins" is that they create a thin hard cap on top of a relatively soft surface, which can be easily damaged under pressure. On surfaces where pressure is constantly being applied—as on an earthen floor—the right choice of sealant would be one that penetrates deeply into the earth rather than forming a shell-like surface. This is why, when sealing an earthen floor with linseed oil, it helps to heat the oil to make it soak in as deeply as possible. To encourage the oil to penetrate even deeper into the floor, it can be thinned with various thinners (see "Earthen Floors" in chapter 7). Good sealants include penetrating oils such as linseed, hemp, caster, or coconut, and animal urine and blood, all of which oxidize and harden the surface.

Certain sealants may be mixed and applied as one layer, but different sealants should never be layered over one another, as they could be incompatible and peel. Binders such as clay can also be used as a waterproof sealant in their finer forms, but too much pure binder will not be durable. Filler (sand, silt, or gravel) should be added to it to make it solid and hard.

Sodium silicate dissolved in water is known as water glass. Water glass can be used as a sealant with certain soil types, but has been known to react very differently

Interior of a *cassita* in Canelo, Arizona. The walls have a clay finish with mica for shine. The white bench is plastered with gypsum.

Cedar Rose demonstrating the spraying of a mud mixture at the Natural Building Colloquium, Kingston, New Mexico.

The Steens' residence, Canelo, Arizona.

with varied mixes of earth. It is not suitable for clay soils, but has proved useful with sandy soils. Sodium silicate is fairly cheap and is available in many parts of the world. It acts as an impermeablizing agent after a curing period of seven days. It is soluble in water, but can be rendered insoluble by allowing it to react with slaked lime. Thin sodium silicate with water prior to mixing with earth, otherwise too many "micro fissures" will result, causing a strong suction of water. Another silicate used to make plasters impermeable is *potassium silicate*. Potassium silicate can be dissolved in water to make a liquid, then used to seal and waterproof mud plaster. The coating is clear in color, allowing the full beauty of the plaster to show. When put on top of lime plaster, potassium silicate reacts with the calcium in the lime and the carbon dioxide in the air, creating an impervious layer. Potassium silicate is made out of quartz sand and potash and binds itself chemically with silica when applied, which is good for surfaces that contain sand. Silicates also bind mechanically in grooves in the surface of the plaster.

Before applying, thin silicates with water: for example, 1 part potassium silicate to 5 parts water.

PAINTS

As well as plastering earthen walls on the inside, you can coat them with "breathing" paints to give extra protection or color. I learned about this subject from Swiss painter Reto Messner. Natural paints derived from plant and mineral materials have subtle colors, pleasant scents, and help create a healthy indoor environment. Since only a few decades ago, the petrochemical industry has largely taken over production of oil-based and water-based paints. They not only abandoned the traditional view of paints as a breathing skin, but have also introduced synthetic chemicals that can be very harmful to us. In addition to being concerned about the pollution generated by industrial paint manufacturing, increasing numbers of people find themselves affected by the vapors given off by modern paints as they dry, and even by the low levels of volatile organic compounds (VOCs) that continue to outgas afterward. Sick building syndrome has now been widely recognized to be the result of modern synthetic materials in combination with poor ventilation; this phenomenon can damage the health or at least affect the comfort and performance of people living and working in newer buildings.

Like cement, modern waterproof masonry paints, emulsion paints, and vinyl wallpapers all slow down or prevent the evaporation of moisture from the wall, which often leads to the render or paint separating from the surface of the wall, and water trapped between the wall and the paint can be harmful to the wall itself, especially if the building is constructed of materials such as wood or earth that can deteriorate when wet.

Paint consists of *pigments, extenders* (also known as *fillers*), and *binders*. Pigments and extenders comprise 75 percent of the total quantity of paint. Binders comprise 25 percent. Pigments color the filler. They can be different minerals or plant powders. It is better to use water-based paints on walls, because these allow for vapor diffusion and breathability. Extenders are a kind of paint "filler." Extenders can include whiting, obtainable at paint or ceramic stores; barium sulfate; kaolin; marble dust; chalk; and diatomaceous earth.

Binders can be water based or oil based, depending on the requirement. Examples of water-based binders include clay; casein (milk protein), which is very strong when dry and will not come apart; water glass; lime cellulose glue; starch glue (corn starch) or wheat paste; and lime.

Limewash, also called whitewash, is a water-based paint that has been used for centuries, ordinarily re-coated every year or so. Additives to limewash to make it more durable (but less breathable) include linseed oil, tallow; and proteins such as egg white, blood plasma, or casein. See page 97 for a more detailed discussion of lime paints.

Painting with clay paint over an earthen plaster on a straw bale wall in Casa Chika, Kingston, New Mexico.

CLAY SLIP OR *ALIS*

Alis is a clay-based paint traditionally used over earthen plasters on the interior of adobe houses. *Alis* can be of any desired color, and when dry it makes a durable finish. I learned about *alis* from Carol Crews of Gourmet Adobe in Taos, New Mexico.

An earthen plaster shower wall painted with lime, Canelo, Arizona.

If white *alis* is desired, Carol uses white kaolin clay as the binder, because it is inexpensive and can be purchased in large bags from a pottery supply store. She uses ground mica in a fine powder form or fine sand as the extender, and straw or mica flakes for added texture and glitter. To make the paint thicker, especially in the first coat, a small amount of fine sand is added to smooth out irregularities in the plaster surface. As filler, Carol uses cooked flour paste in a proportion of 20 to 25 percent of the liquid. Pigments or colored clay can be added to white *alis* to give it color.

To Cook Wheat Flour Paste

Set a pot two-thirds full of water to boil on the stove. In a mixing bowl, whisk together another one-third proportion of cold water with some flour to a consistency of a pancake batter. When the pot of water is vigorously boiling, pour in the flour-water mixture to make the pot almost full, and stir the mixture well.

It should thicken immediately and become translucent. Take it off the heat.

The total proportion of water to flour should be about 6 parts water to 1 part flour paste. For mixing the plaster dilute this wheat paste to a good consistency.

To Make Alis *Clay Paint*

Fill a 5-gallon bucket three-fifths full with 3 parts water to 1 part cooked flour paste.

For the first, slightly thicker coat, add to the diluted flour paste liquid a mix of 3 parts clay to 2 parts mica to 1 part fine sand. (If no mica is available, substitute fine sand.) Keep adding these proportions until the mixture is the consistency of heavy cream.

For the second coat, add to the diluted flour paste a mixture of 1 part clay to 1 part mica. Again, keep adding these until the mixture is the consistency of heavy cream.

A little powdered milk (with casein) will thicken the mixture and makes it somewhat tougher. Some finely chopped straw or large flakes of mica can be added for an interesting texture. Colored clays or pigments may also be added to create different colors. If colored clays are added, they can replace some or all of the kaolin in the recipe. If mold is a problem, it is advisable to add some dissolved borax powder, which will make the paint alkaline.

Application

Before application, make sure the wall surface of your earthen plaster is *totally* dry, as any moisture could leave water stains in the finish. Start applying the paint with a brush at the top of the wall so you do not get drips.

Most walls require two coats. Make sure the first coat is completely dry before the second is applied. When the second coat becomes "leather hard," take a damp (squeezed-out) sponge and a bucket with warm water and start to sponge the wall in circular movements. This will smooth out the brush strokes and clean the paint of any pieces of straw or flakes of mica that might have been added to the mix for special effect. When the sponge begins to feel dry, wet it and squeeze it out again. When you are finished, you can save the leftover paint for later repair to any minor damage by drying it out on a tarp in "cookies." To reconstitute, simply add water to the dried paint until the appropriate consistency is obtained (Kennedy 1999, 93).

LIME PAINT OR WHITEWASH

If possible, find a source for ready-made lime putty (matured for a minimum of six weeks), as this is the easiest, safest way to buy the material, or mix bagged hydrated lime. Limewash is lime putty and water (see page 84 for more on lime putty). Mix the putty and water together to a consistency of skimmed milk. Sieve through a fine kitchen sieve. If pigment is desired, be sure that it is thoroughly premixed by adding it to a little warm water in a jam jar, then seal the jar's lid and shake vigorously. Add the pigment to the limewash, and stir and sieve it again. To get a really white finish over an earthen plaster, you will probably need at least three coats. Limewash turns very white only when it is dry, and if pigment has been added, the final coat will turn about seven times lighter than when it is liquid. Always wet the surface of the earthen wall before applying a limewash.

Additives to Limewash

These include:

- casein glues, or flour paste, as binders
- salt, to improve durability
- molasses, to increase penetration into an earthen wall
- alum, to improve adhesion
- linseed oil or tallow, to increase water resistance

Recipes for Water-Resistant Whitewash

Whitewash with oil or tallow. For exterior surfaces, the addition of linseed oil or tallow (animal fat) makes the whitewash more resistant to water. According to "Appropriate Plasters, Renders and Finishes for Cob and Random Stone Walls in Devon," published by Devon Earth Building Association, use no more than 1 tablespoon of linseed oil or tallow for 2 gallons of whitewash. Add the oil when the lime

Natural Putty

Mix chalk and linseed oil, and knead until well stirred and stiff. Add earth or plant pigment as desired, and use for filling gaps and puttying windows.

and water are slightly heated. Continue stirring and heating slightly until the mixture has blended.

Whitewash with linseed oil and milk. To increase the water resistance of whitewash, fill a container with 4 quarts of milk, add 1½ cups of linseed oil, and stir well. Add lime and stir continuously until the mixture is creamy, of a paintlike consistency.

Some Old Limewash Recipes

If you are interested in trying out working with lime in a more primary form instead of premix, here is a recipe for a brilliant whitewash that will not rub off, and which bears a gloss like ivory. Take 5 or 6 quarts clean unslaked lime, slake with hot water in a tub, and cover securely to keep in the steam that's generated by the slaking. When the lime mix is ready, pass it through a fine sieve, and add ¼ pound whiting, 1 pound good, pulverized sugar, and 3 pints rice flour, first made into a thin paste. Boil this mixture well, then dissolve

1 pound clean glue in water, and add this solution to the mixture. You may also add coloring matter to give the mix any shade you please. Apply the limewash while still warm, with a whitewash brush—except when particular neatness is required, in which case use a paintbrush.

This recipe is from a book by Sam Droege first published in 1861 and found on the Internet (see the bibliography).

To make whitewash that will not wear off, make the whitewash in the ordinary manner, but then place it over a fire and bring to a boil. Then stir into each gallon a tablespoon of powdered alum, ½ pint of good flour paste, and ½ pound of glue dissolved in water while it is boiling.

This wash looks as good as paint, is almost as durable as slate, and will last as long as paint. The recipe comes from the Building Biology and Ecology Institute of New Zealand (see the resources list).

Casein

Casein can be used as a binder or stabilizer, to make paints more durable and weather

---

Casein Glue

Here is an old recipe for casein glue.

Mix skimmed-milk curds (quark) well with 2.5 percent quicklime. Use after one hour. Apply to both surfaces to be glued together and set under pressure for twenty-four hours. The mixture should be remain useable for about three days on wood, cork, paper, or heavy wallpaper.

resistant. When it is mixed with an alkaline substance such as borax, it will react and form a gluelike solution that can be used as a binder in paints or plasters, or as an actual glue (see page 98). If the casein paint is used on top of lime plasters it does not need to contain borax, as the alkalinity in the lime will activate the casein.

Casein can be purchased. One commercial brand is called Auro (see the resources list). You can paint your house using about 2 pints of casein, as it goes a long way. You can also make your own casein with simple ingredients.

To make casein (also known as quark or milk curds) at home for use as the binder in paint:

Mix 1 quart nonfat or 2% milk and 2 teaspoons low-fat sour cream. Stir well. Let sit in a warm spot for two days until it thickens or curdles. (If it does not curdle, warm it up or add vinegar to make it curdle.) Separate the curds from the whey by pouring the mixture through cheesecloth placed in a sieve.

What you have left is known as quark, which contains 7 to 12 percent casein, the right proportion for use as binder in the paint. Do not leave this mixture to stand for too many years, as casein loses strength with time.

Recipe for Interior-Exterior Casein
To make a durable casein glaze or paint that can be used inside or outside and not wash off in the rain:

Mix together ¼ cup (2 ounces) casein powder with 3 ounces of warm water, and ideally let sit for two hours or overnight. (If homemade quark is used, it does not need to be mixed with water.) Mix the casein solution with ½ cup (4 ounces) of borax dissolved in ½ cup (4 ounces) of warm water to produce the binder mixture. (Borax is an alkaline soap; do not dump in your garden.) This mixture will become a sticky paste and is also know as casein glue, which will lose strength over time and must be kept refrigerated.

To this casein binder add 8 cups of water and mix until the solution is like pancake batter. This can be used as a durable glaze or paint. (The binder mix when diluted can also be used to mix a durable plaster.) When making transparent paint (a glaze) for use on top of a lime plaster, you do not need to add borax, since the lime will activate the casein.

To make this glaze into an opaque paint, mix pigment with an extender such as chalk or any other inert white powder to make a paste. Then add this mixture to the binder in a ratio of 25 percent binder to 75 percent extender with pigment.

You can also make a water-based casein emulsion that is more water-resistant using quark and lime. Mix a quantity of skimmed-milk curds (quark) with about 20 percent lime to obtain casein glue, then thin with water until it becomes a creamy brew. This can be used for several days. Add 2 to 3 percent linseed oil to increase adhesion and durability, especially for exterior application. In addition, you can also use diluted casein paint as a primer, which will increase durability further—no wiping off after only a few weeks. Another way to improve the adhesion of this paint is to add honey to 100 parts curd, 50 parts water, and 20 parts lime.

Application

If casein paint is sucked into the plaster right away and the desired effect is to be a glaze, it is necessary to prime the wall first with a thin mixture of casein binder and water, and allow it to dry. When the paint is applied the water in the paint should be absorbed, but not immediately. As a priming binder, alum is especially appropriate for gypsum-plastered walls. If the paint continues to dust off, it does not contain enough binder, or the wall is not primed properly. Too much binder can create a glassy surface, which can flake off.

Oil-Based Paints

Linseed oil (from flaxseed) is a good oil to use in oil-based paint. This is a "stand oil," which is an important factor in making oil-based finishes, because it oxidizes—and dries—when exposed to oxygen. According to the Canelo Project's *Earthen Floors* booklet (see the bibliography), in the "old days" oils were left to stand exposed to the air to produce the drying effect. However, they can now be produced by injecting oxygen into the oil. Linseed oil

is reasonably water resistant, but if exposed for a long time to moisture will eventually deteriorate.

To make oil-based casein paint, add oil to a casein binder (see the recipe above) very gradually, as in making mayonnaise. Mix a maximum 25 percent of volume of oil into the mixture; 15 percent is usually good, so test the paint on a sample patch before more oil is added for shine. The mixture will eventually start to get creamy.

To make casein- or oil-based limewash, mix 1 ounce limewash (lime putty mixed with water) and 5 ounces casein powder or 1½ ounces linseed oil.

Paint this limewash onto fresh lime plaster that is not totally dry. You must use thin coats. It is best to use linseed oil for the first coat, followed by casein, because casein limewash is harder than oil limewash.

Maintenance

The annual maintenance requirements for a plastered building depend on the nature of the finishes that were used. If the plaster finish is made of earth, maintenance is required at least every two years. A thin coat of earthen plaster should be reapplied on a dampened wall surface. This is best carried out during the summer months. If earthen plaster has been sealed with a capping of lime plaster, an annual fresh coat of whitewash will maintain the finish well, but this is not absolutely necessary, as well-executed lime plasters will last many years before they need to be repaired.

Remember, cement plasters are very brittle and will tend to crack due to subtle movements of the whole structure, therefore requiring repair. Repairs of cement stuccos and renders will always be visible unless the whole area is repainted.

Papercrete (fibrous cement made of cellulose mixed with cement or lime; see chapter 1) can also be applied as a plaster directly on earthbags. Papercrete is still at its experimental stage, but this seems to be a material that is water-resistant (not waterproof), therefore suited to most climates; insulative; and not as brittle as cement render, therefore it might not crack as much and may be easier to apply and repair.

The wear of the finishes is affected significantly by the materials you choose, the quality of the application, the location of the house, as well as the overall design of the house and whether there are any children or animals around, in which case more house maintenance is generally necessary. Annual replastering does not have to be a chore, but can be turned into a fun social event involving the whole family or even the neighborhood.

7

OTHER INTERIOR WALLS, FLOORS, AND FURNISHINGS:
BUILDING WITH CLAY

Once the structure of an earthbag house is complete, it is possible for the interior of the house to have conventional furnishings, just like any other house. More often, however, those who have gone to the trouble of finding out how to build an earth house are interested in using natural materials and earth-based techniques for completing the interior, in order to create an especially healthy and beautiful home. This can include partition walls, ceiling treatments, insulation, floors, and furniture. In this chapter, I refer to building materials and techniques such as clay-earth mix, straw-clay, cob, and adobe, which were introduced in chapter 1. In describing these techniques, I emphasize the need for clay as a binder, as this is the most crucial ingredient when building elements that must maintain their structural integrity without exterior forms. Clay generally binds together filler materials such as sand, straw, and earth.

The advantage of using earthen materials for furnishings on the inside of a house stems not only from the beauty and health benefits of earth, but also its ability to adapt to any shape. In a circular earthbag dome shape, where straight and square furniture can be difficult to accommodate and expensive to custom-make, earthen materials are more flexible and less expensive, allowing you to design the interior in a complex or simple way, using straight lines or curves as desired.

FINDING AND ANALYZING BUILDING SOILS

As we consider guidelines for soil analysis prior to building with clay-earth mixtures, readers will once again appreciate earthbag construction for its simplicity, as there is no need to understand the complexity of the soil and its clay content, since the bags themselves provide a form to hold the earth in place, whereas when building interior partitions, benches, ovens, and earthen floors, the binder-to-filler proportions are critical, because these structures need to maintain their own form, with no bag or other exterior form to hold the shape.

Due to the vast variety of soils, there are no universal recipes for making good clay-based mixtures such as cob or adobe. You

Facing page: Earth-plastered shelving being constructed with straw-clay blocks and carrizo decking.

can ask the people who have already built in the area where you want to build what mixtures have worked for them. It is crucial to know your soil before using it for construction. For this, homemade tests can be carried out. There are several characteristics to look for, including particle size (ranging from fine silt and sand up to rough aggregate such as gravel), plasticity (the capacity to retain a shape, which permits sculpting of the material), compressibility (to increase adhesion, especially important with techniques such as production of compressed adobe blocks), and acidity or alkalinity (which affects the way various materials combine).

Prior to beginning to build with earth, it is important to understand the best ways of finding, extracting, and mixing the most resilient blends of clay and other materials. The soil should be obtained from below the topsoil line (topsoil is for gardens), and must be free of all organic matter.

There are a number of ways to find clay. Many conclusions may be drawn from the geological situation. Purchase geological maps of the area or visit the geology department of a university or a government institution to request assistance. Consult local brickmakers or potters, who are necessarily very conscious of fine distinctions among earthen materials. Investigate the availability of clay on conventional construction sites, where clay is often dug up during excavation and transported to a dumping ground, which might be far away and therefore incur extra charges. If you can take this "waste" material off a builder's hands, you may be able to obtain clay for a good price. Certain plants indicate the presence of clay soils. For example, the group of plants called horsetail or scouring rush (*Equisetum*) suggest clay soils (Andreson 1997). When the earth is very dry and many irregular cracks have appeared on its surface, this indicates the presence of clay. This is easily seen at the bottom of puddles or dried-up ponds. Other landscape clues for the presence of clay are described in Michael Smith's book, *The Cobber's Companion* (see the bibliography).

The best way to begin to know your soil is by making several tests with samples.

Jar Test

Pour several handfuls of earth into a large glass jar half full of water. Shake the jar well and let it sit until all the particles settle. The heaviest particles, such as rocks and

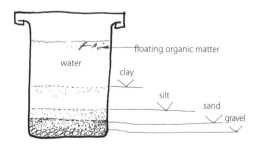

A jar test to estimate percentage of clay in a soil sample.

pebbles, will settle at the very bottom of the jar. Then the sand and the silt (which is a finer version of sand) will settle, leaving the clay (the smallest particles of earth) as the top layer. From this simple test we can estimate the percentage of clay in the soil.

Testing by Hand

There are numerous tests that can be carried out on soils to check if clay is present. Rolling and pressing the clay between the fingers will give an indication if the soil contains any clay. If a thin "sausage" (about ¼ inch or 4 millimeters thick) can be rolled and does not crack very much when slightly bent, the earth contains high quantities of clay. If it cracks, it probably contains a larger quantity of silt. Another way to test soil by hand is to make an egg shape, then crack it. If it resists cracking, it is clay, whereas if it cracks easily it is mainly silt or other more granular particles of soil.

The Right Mix

A good construction material must have clay and sand in the right proportion. The more filler that is mixed with the clay, the more evenly distributed the cracks that result from drying will be. The more the straw is added, less sand may be needed, because straw takes up the shrinkage, therefore stopping the cracks.

Always make samples, since this is one of the most effective ways of being sure of your soil's limitations. A set of samples should be made into small patties or adobe blocks for comparison.

The first sample should be pure earth, to be used as a control, followed by samples with 10, 20, and 30 percent of added sand. Repeat the test, adding fibers such as straw, grass, hair, or textile strands, and then add both fibers and sand in varied proportions. If the soil has a high clay content, you may need to add more sand as a filler or more fiber to inhibit cracking.

When a sample indicates a good mix for building, it will not crack. But not all cracks are bad. For adobe, the California Building Code allows cracks up to 2¾ inches (7 centimeters) long and 1¼ inches (3 centimeters) wide (Khalili 1986). When dry, a promising sample can be tested by twisting it with your hands to try to break it, or by dropping it from knee height. If it does not break upon impact, then the earth mix is right.

If the local soil does not contain enough clay to bind together properly, it is possible that stabilizers are required. These need to be tested at this stage by being added into the mixture in varying quantities.

To answer the question of what makes a good building mix, here are a few pointers, although it is important to note that every mix should first be tested by making small samples and observing them when dry, prior to any application.

- To form a durable surface and create a mix sticky enough to adhere, you may need to mix more thoroughly, increase the clay content, and/or add some type of stabilizer.

- To minimize shrinkage and therefore cracking, you can reduce clay, add more sand or other aggregate, keep water content as low as possible, or add more fiber such as straw and other grasses, cellulose, or other plant or animal fibers.
- To increase water resistance as much as possible and slow down erosion by driving rain, you can add a good distribution of aggregate sizes, fibers such as straw, and/or stabilizers derived from plant, animal, or mineral sources (see chapter 6).
- To increase the permeability and porosity of an earthen mix, which are necessary to permit moisture to evaporate and to allow for expansion of freezing water in order to avoid frost damage, an earthen mix needs a good distribution of aggregate, straw, and anything else that will create air gaps.

If you are carrying out sample tests with stabilizers and are not getting the results you might expect, try testing the pH of the soil, as not all stabilizers react with soils of all types. For example, if the clay is more acidic, it should react beautifully with lime, an alkaline, forming a more neutral and creamy mix for an external render (see chapter 6 for more on stabilization).

You will also want to carefully control the type and size of filler or aggregate that you use, taking into consideration the intended function of the mix—for example, whether it is for the structure of furniture, the filling out or evening out of a base plaster, or the final smoothing out of a finish plaster. The finer the desired appearance, the finer the added aggregate, filler, and fibers need to be.

Also, a more structural earthen mix can have a large range of particle sizes, from silt to gravel, as the filler material, and "filling in," "evening out" plasters can be made using more of the long straw additive in order to be more reinforcing and provide more sculptural capacity. The final layer of plaster can have just fine sand mixed in with the clay, as well as finely chopped straw and stabilizers if desired.

My own favorite mix for a thick first coat of plaster and for sculpting is quite simple and works with most clay soils. In a wheelbarrow (as described on page 108) combine a soupy mixture of clay soil with as much long or chopped straw as the mixture can take and still stick together, along with just enough sand to give it some body and prevent hairline cracks for the finishing layer. (Chopped straw will be easier to manage than long straw during smoothing.) If the first layer is fairly thick and even, the second layer can be thin, with more sand substituted for straw, and inside the house a stabilizer such as wheat paste added.

According to Devon Earth Builders, traditional English cob mix contains clay and aggregates in the following proportions:

fine course sand	25–30%
silt	10–20%
clay	10–25%

It will also contain fibers (as much as the mix can take without ceasing to adhere) to reduce cracking and increase the insulation value. Sufficient straw in the mix provides a level of thermal insulation that is better or equal to the insulation in many conventional houses. In England, traditionally the fibers used were wheat and barley straw along with hay, twigs, and other organic material including animal hair and animal dung.

Correct mixing of the material is as important as the actual construction process. If too little water is added, the necessary distribution of clay throughout the soil will be difficult to achieve, and the cob lumps will be difficult to compact when placed on the wall. In the past, compaction was achieved using the worker's boot, so that each cob is well heeled-in and thoroughly trodden between each course. Excessive moisture dilutes the soil to a porridgy state, making construction impossible; in such cases, more dry mix and/or straw can be added.

Getting the right mix for earthen plaster, floors, or furniture and making adobe

A cob shelter built during a workshop led by Sunray Kelly and Carol Crews, Rico, Colorado.

or cob for building is all principally the same process. The ingredients might vary slightly, but once the recipe for a particular type of earth is established, these proportions can be used for any of the above techniques. The ingredients might include varying proportions of straw, depending on the coarseness of the earth and sand. Earthen plasters or floor finishes will require the mix to be sifted, for a finer finish. Cob for building may require more straw than adobe requires.

Many different ways have been developed throughout the years in different parts of the world, but they all have the same aim—evenly distributing the various materials that make up the mix and creating a moist, pliable mass of earth.

Mixing with bare feet.

In the traditional way developed over centuries in Devon, England, once proportions are identified through testing and making samples, the desired earth mix is

spread out in a bed approximately 100 millimeters in depth on a thin layer of straw. Water is then added and a second, thicker layer of straw is spread evenly on top. (About 25 kg. of straw per cubic meter of soil—1.5 to 2.0% by weight—is considered adequate.) The straw is then trodden into the soil, which is turned several times, more water being added as required. Thorough treading of the mix (traditionally by men or animals) is vital because it ensures even distribution of the clay and renders the material to a consistency and a state of cohesion suitable for building. The quantity of water used will vary according to soil type but is usually in the range of 10 to 12% by weight. If too little water is added the necessary distribution of clay throughout the soil, will be difficult to achieve. (Devon Historic Building Trust, 1992.)

The Taos Pueblo way of mixing is similar to the traditional Devon way, except that the warm sunny summer weather allows the people to do it barefooted in a dug-out shallow pit.

Athena and Bill Steen, while in Mexico working on the Save the Children Foundation project, were taught by their Mexican colleagues how to mix the "no-effort way." Simply half fill the container you are mixing in with water (most likely a wheelbarrow), then use a shovel to sprinkle the dry earthen mix into the water in the proportions you have derived from testing (or use pure clay). Make sure the distribution is even and not too thick. If the mix is not ready, alternate shovels full of the necessary clay, earth, and/or sand. Once earth covers the top of the water, go and have a cup of tea or a lunch break. Let it sit long enough for the soil-clay mixture to absorb all the water. The speed of absorption depends on the fineness of the clay-soil particles (for instance, whether it has been sieved). After a few minutes, test the mix by sticking your finger in. When ready, it will be smooth and creamy. If it is still lumpy and hard, or partly dry, it needs to sit longer. If the soil has not been sieved, up to half an hour might be necessary. When the mix is ready, work it thoroughly with your hands, stirring around so that the sand, clay, and water are well mixed into a soupy consistency. At this stage, as much straw can be added as the mix can take and still cohere, kneading thoroughly with hands or with feet. For foot mixing, dig a pit to use instead of a wheelbarrow.

Another way of mixing, which I learned at the Natural Building Colloquium, involves placing the earth mix in a pile on top of a plastic tarp. Water is added, and two people hold the tarp at opposite ends, leaning and pulling to each side, shifting their grip on the tarp to roll the mixture back and forth, and stopping from time to time to add more water and straw. This technique requires a substantial amount of effort, but, as when mixing with feet, your back remains straight, whereas mixing by hand requires bending over a wheelbarrow. Remember that the thoroughness of the mixing contributes to the binding strength of the resulting mixture.

An endless variety of mixing methods is constantly being developed and refined, each one suiting different climates and individuals. Ianto Evans and Linda Smiley of the Cob Cottage Company, who have devoted the past several years to the revival of cob in the United States, believe that you should always mix cob when happy. This way the building is built with good as opposed to bad energy, enhancing its quality.

Mixing earth usually becomes a much-loved activity, where not only the adults but the children can all join in and have an excuse to get muddy. And it is amazing to see how quickly adults turn into children when working with "mud," especially for the first time. Therefore to replaster your house once a year could turn into a huge party and an excuse to enjoy yourselves with your friends and relatives.

THIN PARTITIONS AND CEILING PANELS

As an alternative to conventional gypsum drywall, extensive research is underway to produce structural members out of clay-fiber composites. Prefabricated fiber composite board is a form of industrial dry-board developed in the past few years (Andreson 1997). This board is made of

Mixing the clay.

fiber-coated, plant-fiber-reinforced clay, manufactured by applying clay to burlap fabric (jute net). For strength, two or more layers of reed mats are inserted crosswise, with alternating layers of clay paste. Finally, the surface of the board is covered in burlap and transported to a drying station. Tests with this kind of board have shown excellent results with regard to fireproof, soundproof, deformation, and diffusion values. Such boards could be used, for example, as a permanent form combined with straw-clay or blown-in cellulose insulation, or as ceiling panels (see Construction Resources in the resources list). They can be screwed, nailed, and sawed, then plastered over as a finish. When a smooth surface such as wood framing has to be plastered, burlap can be placed over a wet coat of base plaster and allowed to dry before a final coat, or reed mats can be used as lathe between sections of clay board. And as noted in chapter 1, interior partitions can be made with wattle and daub or with the rammed straw technique described below.

INSULATION

Good insulation in a building can significantly reduce the heating cost. Insulation can be built into the walls, ceiling, roof, and in some cooler climates the floor (see the section on floors, page 115). It is a material that will trap small pockets of air. Natural alternatives to industrial fiberglass include straw treated with potato starch or

borax for fireproofing or stuffed into treated burlap bags, cotton, flax, or sheep's wool. Cellulose insulation made out of recycled paper ground up and treated with a flame retardant can be blown into cavities between rafters with a specially rented machine to insulate the ceiling or roof, and in timber frame construction, between studs in the walls (it is also available in batt form). Hemp cellulose (fireproofed with mineral salts and called Canobite) can also be purchased, either loosely packed in bags or to be blown in. Wood-fiber boards can also be used for thermal and acoustic insulation (Tibbles 1997–98).

To increase insulation value in an earthen mix, straw, wood fibers, cork, and other air-trapping fibers can be used, either added into the earthen mix or attached in panels along the walls. Pumice, perlite, and other minerals used for floors and screeds can also be added as an aggregate to increase the insulating value of interior plaster and ovens. One of the best insulating materials for corbeled dome construction is pumice-filled bags (see the profile of Kelly and Rosanna's house on page 135), but this material occurs naturally in very few parts of the world and is costly to purchase and transport.

Earthbag domes in sunny climates do not require insulation, if the design provides for passive solar heat in the winter, and openings are placed in a way that they do not allow direct sunlight into the house in the summer. If the dome is lived in and heated regularly either by the sun or a stove, the thermal mass of the earth will retain heat in the winter and coolness in the summer.

For earthbag domes built in cold climates, if desired the insulation can be incorporated in the floor, ceiling, and final layers of the internal and external plaster, which can be significantly thickened to provide a more insulative finish using a layer of straw-clay on the inside and papercrete outside.

An earthship (rammed earth in tires) wrapped by straw bale insulation.

Straw–Light Clay

As mentioned above, straw can be coated with clay slip called straw–light clay and used as insulation in many different ways: stuffed between rafters as roof or ceiling insulation, placed in the floor using the rammed straw technique, or made into lightweight blocks to construct relatively thin but highly insulative walls.

To make light clay, pure clay is necessary for maximum binding strength. The clay has to be mixed with enough water to turn it into a slurry called clay slip or liquid clay (always adding the clay to the water, never water to the clay). If the mix is lumpy or contains stones, it could be passed through a ⅛-inch screen. The consistency should be such that when you dip the palm of your hand in the mix no lines can be seen on your hand. The purer the clay, the thinner it can be diluted due to its greater binding strength, thereby achieving a lighter straw-clay mix.

To prepare straw–light clay, pour the clay slip on top of a pile of straw, tossing it like a salad with pitchforks. Coat every single piece of straw completely with clay slip. To test if it is coated enough, take a bundle of the mix and squeeze it—if it sticks together, it is ready. If time allows, it is then best left for a day or two under a tarp to mature and improve.

Another application for light clay is the technique of *rammed straw*, whereby the mixture described above is used to construct walls and partitions. After being coated with clay slip, the straw–light clay is lightly rammed between the form boards (shuttering) with a 2 x 4 (or your feet) until it is solid and not spongy. The forms can be moved up to the next lift immediately after completion of each particular section. To preserve the insulation properties (that is, trapped air in the straw), it is important for the tamping not to be too hard.

112

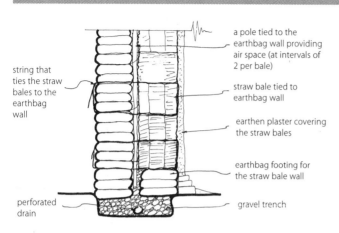

string that ties the straw bales to the earthbag wall

perforated drain

a pole tied to the earthbag wall providing air space (at intervals of 2 per bale)

straw bale tied to earthbag wall

earthen plaster covering the straw bales

earthbag footing for the straw bale wall

gravel trench

Detail section through an earthbag wall with straw bale wall for added insulation.

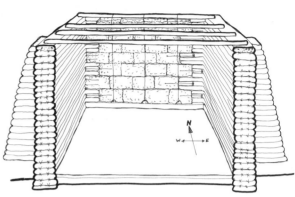

A house where two (east and west) earthbag walls are structural and the straw bales are a nonstructural insulating wall on the north side, with glazed frame construction on the south-facing front for passive solar heating.

Three different ways of integrating earthbag and straw bale wall systems.

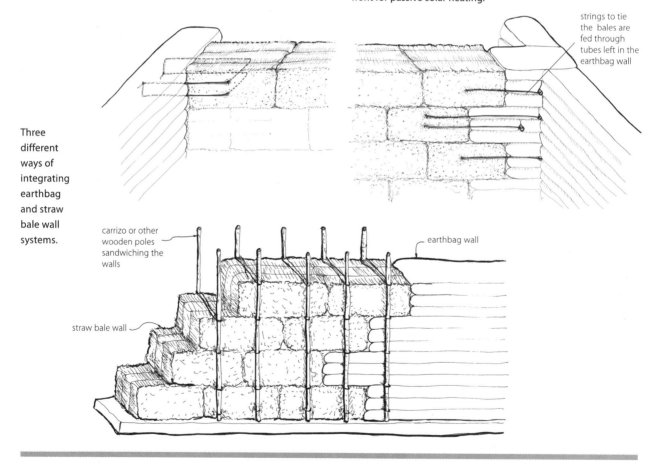

strings to tie the bales are fed through tubes left in the earthbag wall

carrizo or other wooden poles sandwiching the walls

earthbag wall

straw bale wall

Hybrid Earthbag and Straw Bale

Straw bales are among the best value for natural insulation, but unfortunately they take up a lot of space. They can be used as floor or roof insulation on ladder trusses or as insulation for living roofs. Bales can also be used in conjunction with earthbag walls either as internal supplementary insulating walls, creating a three-foot-thick wall, or as nonstructural infill in combination with load-bearing walls or piers, creating buildings that have the structural stability and thermal mass of earth (on the east and west side, for example) and the insulation value of straw bales on the cold north side.

Straw is an annually renewable resource, the waste product of a cereal grain crop, and can be easily grown and harvested. Bales can also be made out of tumbleweed, sudan grass, and ordinary meadow hay, but straw is the best natural insulator due to its hollow stems that trap the air, and it is not attractive to vermin since it lacks nutrients. The straw bale technique represents an entirely distinct construction system, which needs to be

Above: Building with rammed earth.

Left: A rammed straw wall.

understood as a system in itself before it can be properly applied in combination with earthbags, adobe, or cob. Straw bale structures can be load bearing or non-load-bearing. A number of good books are available on straw bale construction (see the bibliography).

INTERIOR DETAILING

Several earthbuilding techniques can be used for sculpting furniture as well as for construction of small structures and houses. Earthbags, rubble, cob, adobe, the different straw-clay mixes, and straw bales can all be used to construct the main structure of sculpted furniture, sealed and smoothed out with an earthen plaster, then capped with lime, gypsum, or some other clay paint finish for durability.

The materials used for creating furni-

ture can also be used for stoves and ovens by adding less straw and more sand, perlite, or pumice to the mixture. These earthen heaters can be sculpted with niches, alcoves, and benches to suit the size and shape of a house. By embedding a flue in an earthen bench, you can make a warm seating area.

For the interior partitions or even exterior walls where the earth has insufficient binding strength to hold nails carrying the weight of fixtures, the installation of hanging cabinets and other furnishings requires a nailer board that can be installed after the wall has dried. In preparation, wooden stakes should be placed between the rows of earthbags before tamping, or embedded in the cob or adobe with a ledger board attached across two or more stakes, which provide anchors for nailing.

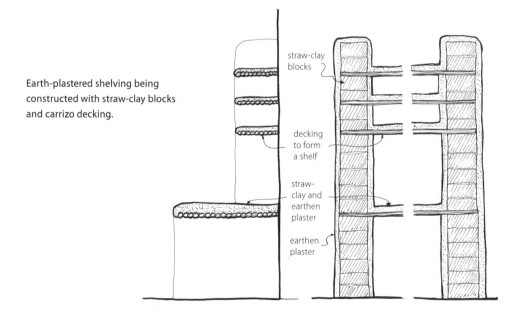

Earth-plastered shelving being constructed with straw-clay blocks and carrizo decking.

straw-clay blocks

decking to form a shelf

straw-clay and earthen plaster

earthen plaster

In addition to being used in wall construction, cob can be used to sculpt furniture in a very freeform way—seating, desks, and shelves. Benches can be made of solid cob with a flue from a woodstove coming through and warming them while the stove is used, creating a "cozy" corner through the winter months.

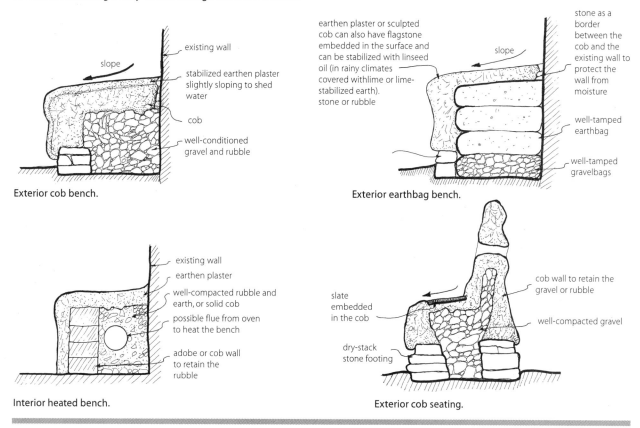

Exterior cob bench.

existing wall

stabilized earthen plaster slightly sloping to shed water

cob

well-conditioned gravel and rubble

slope

earthen plaster or sculpted cob can also have flagstone embedded in the surface and can be stabilized with linseed oil (in rainy climates covered withlime or lime-stabilized earth).
stone or rubble

slope

stone as a border between the cob and the existing wall to protect the wall from moisture

well-tamped earthbag

well-tamped gravelbags

Exterior earthbag bench.

existing wall

earthen plaster

well-compacted rubble and earth, or solid cob

possible flue from oven to heat the bench

adobe or cob wall to retain the rubble

Interior heated bench.

slate embedded in the cob

dry-stack stone footing

cob wall to retain the gravel or rubble

well-compacted gravel

Exterior cob seating.

Then a cabinet or other fixture can be attached with screws or nails to this ledger board.

EARTHEN FLOORS

For centuries, earthen floors were used all over the world. Until recently, they were the standard floors throughout the southwestern United States, where they are currently being revived. In spite of the stereotype of "dirt floors," earthen floors need not be dusty, fragile, or difficult to clean. The right application of oil and wax makes earthen floors waterproof and almost as durable as concrete. Some of the earthen floors I have seen have been walked on with high heels, and can take the pressure of furniture. There is no one correct way of constructing these floors. The materials used vary according to availability, but the quality of the floor primarily depends on the workmanship. It is possible to create earthen floors that are durable and require little maintenance, if certain basic principles are understood. The method I learned from Bill and Athena Steen of the Canelo Project, who have produced an introductory earthen floor booklet (see the bibliography); therefore, I will only give an outline of the construction method.

The floor should be poured during the driest part of the year. It can take between ten days and six weeks to thoroughly dry out. The layers of an earthen floor are as follows (make sure each layer is fairly level to minimize your work on the final layer):

Layers of an Earthen Floor

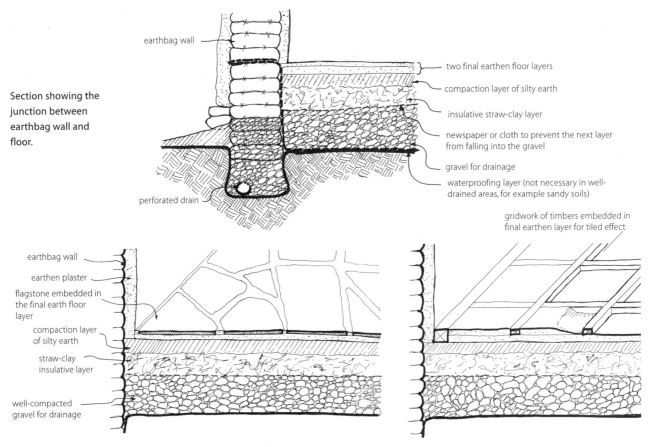

Section showing the junction between earthbag wall and floor.

earthbag wall

two final earthen floor layers

compaction layer of silty earth

insulative straw-clay layer

newspaper or cloth to prevent the next layer from falling into the gravel

gravel for drainage

waterproofing layer (not necessary in well-drained areas, for example sandy soils)

perforated drain

gridwork of timbers embedded in final earthen layer for tiled effect

earthbag wall

earthen plaster

flagstone embedded in the final earth floor layer

compaction layer of silty earth

straw-clay insulative layer

well-compacted gravel for drainage

Earthen floor layers with flagstone embedded into the top layer.

1. Base: This layer should be either undisturbed soil, or at least a very well-compacted surface, since it needs to be free of all organic matter and unlikely to heave in frost, so that no movement occurs.

2. Waterproofing: This layer is only necessary in very damp areas that cannot be properly drained. The waterproofing can be natural clay such as bentonite (which should be tested prior to construction) or a synthetic damp-proof membrane.

3. Drainage: To stop any moisture from rising, this is a layer of 6 to 12 inches (150 to 300 millimeters) of washed gravel or course sand, tamped down well. If no waterproofing membrane is used, the gravel needs to be quite large—¾ to 1½ inches (20 to 40 millimeters) in diameter—to prevent the rise of moisture.

4. Insulation: This layer should be 4 to 6 inches (100 to 150 millimeters) of straw–light clay or pumice, perlite-clay, or bottles embedded in sand, well tamped.

5. Subfloor: The aim of this layer is to achieve maximum compaction on top of the insulation, in preparation for the finishing layers of earth. It should be compacted silty or sandy soil, the same as the base. This is the layer that can take radiant-floor tubing. If a floor heating system is used, the subfloor must be considerably thicker to provide adequate thermal mass.

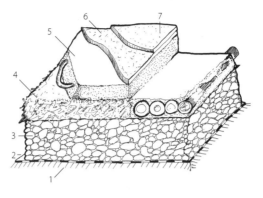

Floor, showing alternative insulation layer of bottles embedded in sand.

6. Finished floor: The top structural layer, approximately 1 inch (25 millimeters) of trowelled clay-earth mix. The clay-earth mix should be comparable to a good earthen plaster mix (see chapter 6), and should be troweled in two half layers; the top layer will need stabilizer. Other options include 4 to 6 inches (100 to 150 millimeters) of well-tamped clay-sand-soil mix or 2 to 3 inches (50 to 75 millimeters) of clay-soil mix with psyllium (the mucilaginous powdered seed of the *Plantago psyllium* plant, also used as a laxative). For natural stabilizers, hardening agents such as lime, blood, or wheat paste can be added (see the discussion of stabilizers in chapter 6).

7. Sealant: Apply several layers of an oil-solvent solution for added protection. Use boiled linseed oil or other stand oil, as these are oils that dry well, and which can also be used for other earthen finishes. For the solvent, the least expensive is turpentine, but you may use anything from common mineral spirits to more-expensive odorless turpentine or pure citrus oils.

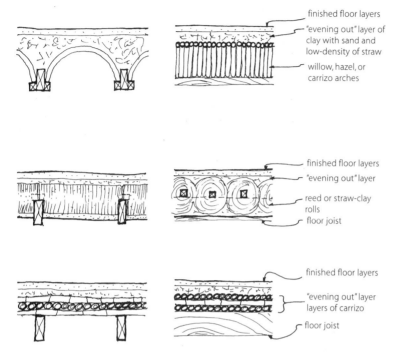
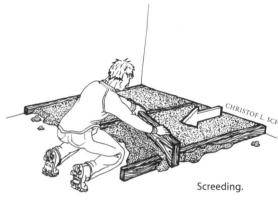

118

finished floor layers

"evening out" layer of clay with sand and low-density of straw

willow, hazel, or carrizo arches

finished floor layers

"evening out" layer

reed or straw-clay rolls

floor joist

finished floor layers

"evening out" layer

layers of carrizo

floor joist

Examples of an upper floor with an earthen floor finish.

screeds. Pour the mix between them, and lay a straight board across the top of them. Remove the board farthest from you, and fill in the void where the board had been with more mixture. Reposition the first board, level it, and keep going. It is best if your mixture is not too wet and keeps its shape when each board is removed. For the final layer, hammer in nails so their heads are level with the height of the finished floor. Use them for leveling the board, and pull them out as you go along.

Screeding.

If you wish to consider construction of an earthen floor for an upper story, a straw-clay mixture can be used to fill in the space between the ceiling joists, as shown above. Then the layers of earth can be poured upon that base as described below.

Construction

Use screed boards as wide as the depth of the layered floor, initially placing them flat near the wall where you will start pouring. Remember to start at the farthest corner, working your way out toward the door. With an accurate level, keep checking that each board is level as you use them as

Here are some other ideas for natural floors. Lay a gridwork of equal-sized 2 x 4 timbers before pouring the final layers, then fill the spaces between them one by one, leveling at the same time. The timbers can be permanent or else removable when the mixture is "leather-hard," and the voids can then be grouted with a different-colored mix. An alternative to a poured adobe floor could be sundried adobe bricks set in place like tiles with a mud-sand mortar. Fired brick, tile, or flagstones can also be set with adobe mortar for areas of the house that get wet frequently, such as the kitchen, bathroom, mud room, or entrance hall.

Sealants, Maintenance, and Repair

The top layer should be completely dry before applying the sealant. To be most effective, the oil-solvent solution should be heated, taking care not to reach the point where it begins to smoke. Warming encourages deeper penetration of the oil into the floor. If the ambient temperature of the room with the floor is warm, the sealant will be better absorbed. Apply with a brush, and remove the excess. Each coat should be applied only thick enough that it does not begin to puddle, for if allowed to puddle it may form a skin on the surface, which will be prone to cracking. Note that both oils and solvents are very flammable and should be treated with caution when heating. Any brushes or rags used during application should be stored carefully in closed containers to prevent spontaneous combustion.

It is better to apply the oil in a stronger concentration in the initial coats, gradually reducing the proportion of oil to solvents in the following coats. The earthen floor is less porous with each subsequent coat of sealant, but will accept full-strength oil at the beginning.

According to the Steens' *Earthen Floors* booklet, the sealant coats can be diluted as follows:

Beeswax for Floors

In a double-boiler with water in the bottom section, melt 24 percent beeswax and 6 percent carnauba wax at 140 to 158 degrees Fahrenheit (60 to 70 degrees Celsius).

Add 30 percent balsam turpentine from spruce or larch and 40 percent boiled linseed oil.

Increase linseed oil and reduce beeswax to make a softer mix. You may use citrus oil instead of pine turpentine.

coat 1—apply full-strength oil
coat 2—dilute the oil with 25% solvent
coat 3—dilute with 50% solvent
coat 4—dilute with 75% solvent

Each coat should be applied only after the previous one is dry. The floor should only need four coats to be sealed. For additional sheen and durability, another coat can be applied periodically. The frequency of maintenance will depend on the wear the floor receives, but for an average floor 6- to 12-month intervals is sufficient. If sheen is not important, the floor does not have to be recoated for many years.

If you want the floor to be not just water resistant but waterproof, after the floor has dried from the last application of oil, apply a coat of wax. Make a paste by melting 1 part beeswax with 2 parts boiled linseed oil. While the paste is still warm, rub it into the floor with a clean rag. The wax layer will rub off over time, so reapply it every few months or once a year.

If cracks or other wear-and-tear begin to show on the floor surface, it is good to patch these places fairly quickly to avoid them growing larger in size. When constructing the floor, save some of the mix for later repairs, as it will be almost impossible to match the color later. After clearing the area that needs repair and crackup off all loose material, add some water to the dry mixture and mix thoroughly. Then wet the damaged area, fill it in with the mix, and reseal the surface as described above, with four layers of sealant.

ELECTRICITY AND PLUMBING

For electrical and plumbing utilities, it is safest to place all the service ducts that enter and leave the house below ground level going through the foundations, to prevent freezing and to minimize damage if a pipe does fracture. This can be planned into the design ahead of time, enabling the insertion of plastic sleeves through the foundation and floor where necessary during construction.

A plumbing chase can also be created in partition walls and under the floor. Non-pressurized drainpipes are safe to route through earthbag walls and can go directly out to a separate graywater system for each sink. If pipes must be run through earth-bag walls, they should be encased in a larger-diameter sleeve that is sloped down to the outside. That way if the inner pipe gets a leak, the outer sleeve will direct the water outside the house where it can be seen, instead of soaking the inside of the wall without anyone knowing.

Electrical wiring and J boxes can be placed as the rows of the earthbags go up, or can be added in the grooves between courses before plastering. Cut 12-inch-long (130 centimeter) pointed stakes out of 2 x 4s to anchor the electrical boxes in the earthbag walls. After making a notch for the box to recess into the earthbag surface, drive the stake in. Even easier, place the stake in the wall between courses during the laying of the bags. You can screw the box to the end of the stake and place the wires between the courses of bag. Use heavy-gauge, U-shaped wire pins to hold the wiring in the grooves. If you are using cement plaster internally, you may want to

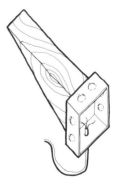

Electrical box fixed to wooden stake for anchoring in earthbag wall.

cover the wires with a thin strip of metal or plastic to keep the stucco from contacting the wires. Wiring that is plastered into a wall is difficult to modify, so test your wiring fully before plastering; or route your wiring in conduit, which can be made out of plastic tubing or discarded garden hose, either set into grooves between courses or exposed on the wall surface for future accessibility and convenient servicing. The other option is to consolidate all the wiring on the interior partitions (if they exist), in the ceiling, or in raised floors with floor outlets.

8

THE EARTHBAG ADVENTURE

Since the inception of this book, numerous earthbag projects have been built. This chapter offers a survey of the earliest and therefore some of the most adventurous. These include examples of the Hart's very experimental freestyle dwelling; the amazing demonstration of courage by Shirley Tassencourt, then in her late sixties; and Kaki Hunter and Doni Kiffmeyer's advancement of the earthbag technique to true perfection. Each of these projects yields tremendous inspiration and many lessons.

SHIRLEY TASSENCOURT'S DOMES, ARIZONA

To my knowledge, the first earthbag domes to be actually inhabited were built by Shirley Tassencourt with help from friends and relatives, including her grandson Dominic Howes. Shirley chose the earthbag technology because of its "magic" and her limited finances. As she is an artist who often sculpts with clay, earth seemed like a familiar medium.

Between 1995 and 1997, Shirley built three earthbag structures: first, a meditation dome, "Domosophia"; next, a main house dome; then, a rectangular-shaped library with a conventional roof. The soil used to fill the bags was brought from nearby.

The Meditation Dome

The meditation dome has an external diameter of 15 feet (4.5 meters)with 1½-foot-wide (45 centimeter) walls sitting on bedrock, a cement-stabilized row of earthbags below the rafters of the mezzanine level, and a reinforced concrete bond beam pinned to the wall, serving as the compression ring for the opening. Exterior and interior plasters are cement stucco. The upper level was

Interior of the main house dome.

Shirley sculpting.

Facing page: Earthbag country.

Section of the
Meditation dome.

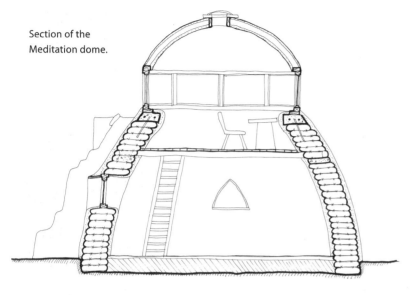

DOMINIC HOWES

SHIRLEY TASSENCOURT

Above: The first lived-in earthbag dome built in Arizona by Shirley Tassencourt with her grandson Dominic Howes.

designed to contain a small studio space with a 360-degree view of the surrounding land.

Here's how Shirley tells the story: "At the age of fifty-two, as a first-time contractor/owner-builder of a 1,500-square-foot saltbox on Martha's Vineyard, I thought I had just made it under the doddering line. But as a retired art teacher and potter/sculptor, I had another go at it in Arizona's high desert. Then as I sat in the dome at Cal-Earth listening to Nader Khalili, I was obsessed with wanting to be in such a centered, dynamic, revolving space. Being old enough (sixty-nine) not to be hampered by reality, I went home and started the next day on the Meditation Dome. Some adventure, oh my! Lady luck hovered over the total project—undertaken by me in my late sixties, my nineteen-year-old grand-

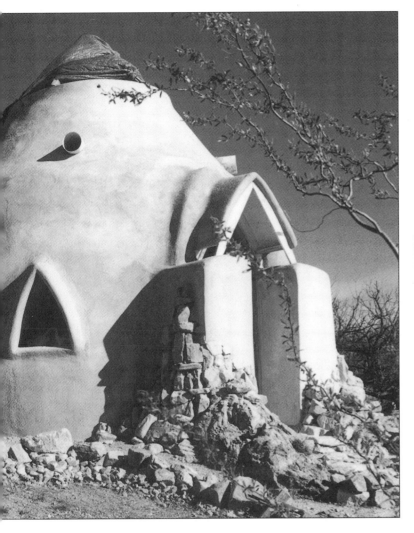

SHIRLEY TASSENCOURT

SHIRLEY TASSENCOURT

son Dominic Howes, and a fun young married couple, Luther and Cindy McCurtis. We figured it out as we went along. With four of us working for five hours a day in the desert heat using small bags, in three months we finished the essential dome, *Domosophia*. A carpentry crew made the 7-foot-diameter clerestory (with its 360-degree view); they brought the spidery structure out in a truck and plopped it on top. We covered it with chicken wire, tarpaper, and stucco, and *voilà*—I had my heart's desire."

Above: The skylight being constructed on top of the compression ring.

Above left: The cost of materials for the main dome structure and finishes (without utilities) was about $6,000, primarily for cement and specially made windows, doors, and skylight.

Insert: The library was built under a pole supported roof, which provided a cover for shade. Earthbags were used as infill, as in a post-and-beam construction.

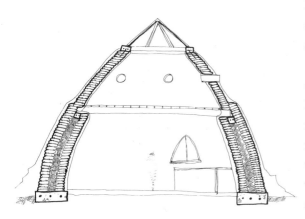

Section and plan of Shirley Tassencourt's Main House dome. Note the double earthbag wall up to the mezzanine level.

The Main House Dome

The second-story floor rests on a concrete bond beam, and the mezzanine level has a sky view. No other foundation was needed, as the ground is bedrock. The earth was dug out 6 inches (150 millimeters), and a poured concrete slab finished with tiles. External and internal plasters are cement stucco without lath; the uneven surface of bags provides enough reinforcement and key-in points.

Again, here is Shirley's own account: "Emboldened by our success, in 1995 in a hot September desert, we started on the second, larger dome, this time built with an engineer's approval. It took us five months and 5,000 small sand bags. This dome has a 25-foot footprint, with a double (42-inch-thick) earthbag wall 9 feet up to the base of the mezzanine level with earth rammed in the cavity, which is sealed with a concrete bond beam. The outer wall acts as a huge buttress, and could have been considerably smaller, but there was little precedent for this kind of construction using small bags. We hand-lifted 25,000 pounds of earth on straw-bale scaffolding for our 20-foot-high building, crowned with a 5-foot-high, 7-foot-diameter plexiglass skylight. Skylights offer gifts of sky and landscape unusual to dome construction. Twelve-inch PVC tubes through the second-floor walls plus a window-door to the balcony allow inexpensive fenestration and continuous air flow from two doors below, which are open all summer. [Author's note: In Arizona, this skylight has to be covered in the summer due to the intensity of the sun, and in winter it causes considerable heat loss. The skylight should be off-center, angled south.]

"My grandson was an apprentice for the first dome, foreman on the second dome, and a contractor on our neighbor Allegra's house (see page 127). Dominic went on to build a large, rectangular earthbag- and roof-truss hybrid structure in Wisconsin. Here in Arizona, I and two other elder women have thirty acres off-the-grid. We embrace Permaculture, gardening till the grasshopper plague arrives in July. We do ceremonies, and connect deeply with the land through our fifteen-acre natural medicine wheel (made with big boulders marking the cardinal directions on a circle). We want to encourage others. . . . If we can do this, anyone can!"

Allegra Ahlquist's House, Arizona

This house is situated on the land shared with Shirley Tassencourt in Arizona. It was built by Dominic Howes, finished in 1997. It is an example of mixing alternative technology with conventional construction. Buttressed earthbag walls stand on a concrete foundation, and are tied together with a bond beam supporting a timber-trussed roof. Allegra is very content with her 625-square-foot house. The house took four months to build and cost 40,000 dollars, the biggest expenses being a conventional roof, foam insulation, concrete foundation, windows, doors, cement stucco, and labor, which was about one-quarter of the final cost. The floor is brick on sand with floor-heating tubing in the sand layer. The south-facing windows provide passive solar heating; in fact, the in-floor heating system (regulated by a thermostat) has only been used six times in the past three years in spite of cold winters.

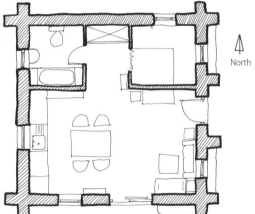

North

Above: Allegra's square house.

Left: Sketch plan of the house.

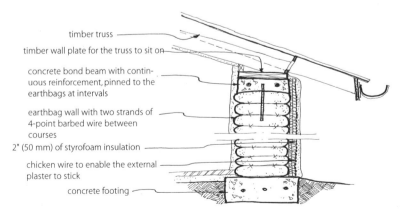

timber truss

timber wall plate for the truss to sit on

concrete bond beam with continuous reinforcement, pinned to the earthbags at intervals

earthbag wall with two strands of 4-point barbed wire between courses

2" (50 mm) of styrofoam insulation

chicken wire to enable the external plaster to stick

concrete footing

Section showing wall-to-roof junction.

House Built by Dominic Howes, Wisconsin

Above: The finished house.

The house is situated in an extreme climate where the temperatures range from as low as −40 degrees Fahrenheit in the winter up to the high 80s in the summer, with heavy rainfall through the summer months. Designed by the client, it has two stories where the second is of conventional construction, with high windows for passive solar access.

This house was built by Dominic Howes almost directly after construction of Allegra Ahlquist's house (page 127); therefore, the treatment of the earthbag wall is carried out in a similar manner. It is almost entirely a conventional house with standard timber-frame construction and artificial foam insulation, but instead of using brick or concrete for the structural walls it uses the earthbags filled with the local earth.

The earthbag part of the house took three weeks to build. The floor area of the house including the first floor is approximately 1,500 square feet.

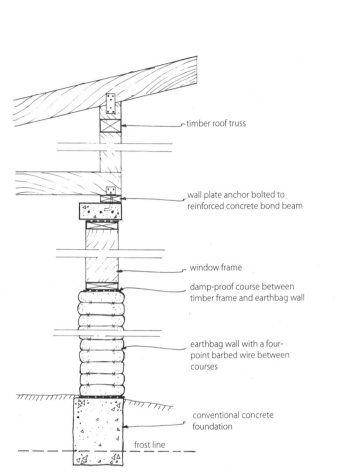

timber roof truss

wall plate anchor bolted to reinforced concrete bond beam

window frame

damp-proof course between timber frame and earthbag wall

earthbag wall with a four-point barbed wire between courses

conventional concrete foundation

frost line

Structural section through the south wall.

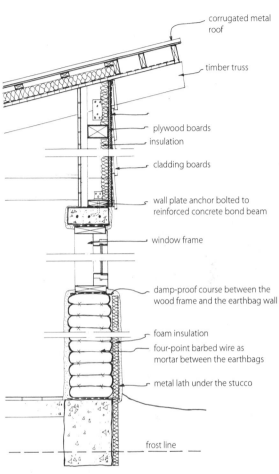

corrugated metal roof

timber truss

plywood boards

insulation

cladding boards

wall plate anchor bolted to reinforced concrete bond beam

window frame

damp-proof course between the wood frame and the earthbag wall

foam insulation

four-point barbed wire as mortar between the earthbags

metal lath under the stucco

frost line

Detail section through the south wall.

The house under construction.

DOMINIC HOWES

DOMINIC HOWES

Sue Vaughan in front of her scoria-filled earth-bag dome, which is covered with standard cement stucco.

Above: The upper section of the dome from the inside, showing the geodesic structure.

SUE VAUGHAN'S HOUSE, COLORADO

Sue Vaughan wanted a very small, round house, so with the advice of Kelly Hart, she and two helpers built this 14-foot (4.2 meters) diameter, shallow dome with a sleeping loft made of scoria-filled bags. A concrete bond beam at the height of the door lintel supports a geodesic structure forming the upper part of the dome.

timber geodesic dome serving as a permanent form

scoria-filled bags supported by the geodesic dome acting only as insulation and thermal mass

mezzanine sleeping level

2" x 6" (50 x 150 mm) continuous bond beam

cement stucco

metal barrel to form opening

square sheet of glass over a round opening

damp-proof course

structural wall of bags filled with scoria

Section of Sue Vaughan's dome.

SUSTAINABLE SYSTEMS SUPPORT

CAROL ESCOTT AND STEVE KEMBLE'S HOUSE, THE BAHAMAS

Here is a house designed and built in the Bahamas by Carol Escott and Steve Kemble of Sustainable Systems Support. The first phase of the construction, which consisted of the structural earthbag walls, was constructed with the expert help of Kaki Hunter and Doni Kiffmeyer, who have developed site-built hand tools and a process for building that simplified and "neatened" this very labor-intensive construction method. Kaki and Doni's Honey House is described on page 140.

This is a hybrid design where the first story is constructed out of bags filled with native soil, upon which sits a second floor and roof constructed out of conventional timber frame. The second-floor joists are fixed to a reinforced concrete bond beam on top of the earthbag wall.

The house was built on a small island very close to the beach. The fill for the earthbags was locally available sand dredged from the sea, which was very fine and contained a high proportion of crushed coral, so it was very easy to compact. When slightly moistened and tamped, the bags turned to solid blocks.

The finished house.

Preparing the foundation.

Since the ground around the house was sand, drainage was not a problem; therefore, the foundation was very shallow with no gravel trench below, simply one row of sand-filled bags below ground level.

In an article in *Earth Quarterly* (see the bibliography), Carol and Steve describe the premise of their design process: "Faced with the challenge of building on a remote island in the Bahamas, we realized that . . . current construction trends in this part of the world rely on the importation of almost all building materials in even the remotest of locations. In keeping with our work in the States, we wanted . . . to use this project as a demonstration of appropriate earth building techniques . . . in hope of influencing a shift in . . . building/development needs."

Carol and Steve faced serious difficulties: The Bahamas are subject to devastating hurricanes each summer and fall. There are voracious termites, making any wood product is subject to attack. It also rains frequently and things may stay moist for weeks. The intense summer sun can cause even pressure-treated wood, if exposed, to deteriorate in as little time as five years.

According to Carol and Steve, the Bahamas have developed very little industry except for tourism, so most building materials are imported from the United States. In terms of native resources, the majority of vegetation is low bushes, with trees for lumber being virtually nonexistent. The most abundant and easily gathered natural building resource is sand. Carol and Steve realized that where dredging had occurred for a marina there were piles of sand mixed with crushed coral, available free for the taking. When slightly dampened and well tamped, the lime in the coral acts as a natural binder with the sand, which sets into a hard block.

Since the climate is subtropical, hot and humid, ocean breezes are needed for comfort. Most people clear-cut the bush to allow the breeze to blow through and to eliminate hiding places for the mosquitoes, but Carol and Steve let the foliage grow around their site in order to provide overstory protection for new plantings, shade to help with moisture retention, and

SUSTAINABLE SYSTEMS SUPPORT

First floor under construction.

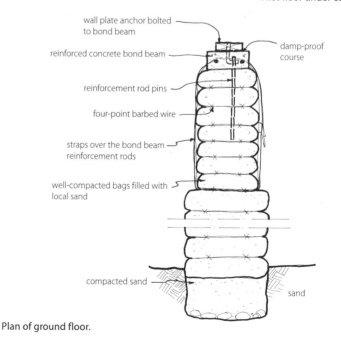

wall plate anchor bolted to bond beam

reinforced concrete bond beam

damp-proof course

reinforcement rod pins

four-point barbed wire

straps over the bond beam reinforcement rods

well-compacted bags filled with local sand

compacted sand

sand

Plan of ground floor.

to encourage biodiversity and habitat for beneficial wildlife, as well as for privacy around the house.

To collect the breezes, Steve designed the home to be two stories with an attached deck on the second level. The roof is hipped to shed heavy hurricane winds. The first level has walls made of sand- and crushed-coral-filled polypropylene bags along the perimeter, with a peaked, 8-point-arch opening in each of the six sides. These arch openings have been left open for breezes, and the covered lower level is used for utility, storage, and a workshop. At the center of the lower level is a 3,000-gallon, concrete-block rainwater cistern, which also serves as a load-bearing support for the second level.

These earthbag walls used misprinted 50-pound rice sacks for the first four feet, resulting in a 20-inch-wide (50-centimeter) wall. The next four feet were built with continuous tubing, resulting in a 14-inch-wide (35 centimeter) wall. Since the earthbag walls were made very smooth with a high-precision finish, there was no rough texture to permit the stucco to be effectively keyed into the wall surface, which to reinforce the cement render was therefore covered with a layer of chicken wire, secured to the walls with galvanized tie-wires laid between bag courses as the walls were built. The bond beam was strapped to the earthbag wall using both the chicken wire and the poly strapping, which was ratchet tightened before stuccoing.

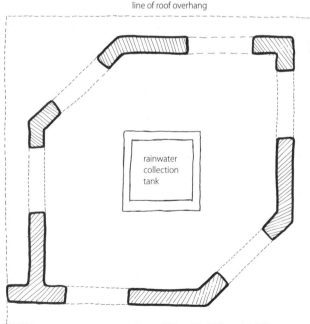

line of roof overhang

rainwater collection tank

Plan of ground floor, used as utility area and workshop.

The 512 square feet (40 square meters) of the ground floor, earthbag stage of the house took three months to build, requiring 500 bags and a 1,400-foot roll of polypropylene tubing.

Carol and Steve conclude: "After the completion of Phase 1 (the earthbag wall base) we were very pleased with the results of the project. The islanders have accepted it, stating that it looks like the old 'rubble stone' ruins around the island. It feels very sturdy . . . enough to take the worst hurricane, relentless sun, and regular wind-blown rains. It cost a fraction of the money any other option would have, to get to this point. Although it took a lot of manual labor, it is a doable method with only minimal hand tools, and we had fun coordinating our team work into a smooth process. The three young men we trained are looking at building houses for themselves using this method, and are even discussing becoming contractors for other people."

Top: Ground floor during construction.

Bottom: Earthbag stairs.

COURTESY OF HARTWORKS, INC.

KELLY AND ROSANA HART'S HOUSE, COLORADO

This house, designed and built by Kelly and Rosanna Hart of Hartworks, Inc., is in a small town at 8,000 feet in the foothills of the Sangre de Cristo Mountains of southern Colorado. Several interesting features distinguish the main dome of this house.

The walls are constructed of bags that contain scoria, a very porous, pumicelike volcanic stone. It is locally available, and like pumice, it has many air pockets and so is very light. Due to its porosity, scoria is a good insulator and also provides thermal mass, so the house can absorb the sun's heat and retain it throughout the night. Kelly estimates that the finished scoria bag

wall covered with papercrete inside and out will have an R-value as high as 40, which is higher than building codes require. Scoria is also light and fast to work with.

Another unusual aspect of this house is that although the main structural form appears to be a dome, it is not a self-supporting corbeled dome. For a true corbeled dome, the design needs to be circular in order to create an evenly curved wall upon which the vertical and horizontal forces are equal everywhere. In this case, the plan is oval, so timber poles have been used as a type of pitched-roof teepee ar-

Roof structure to support the scoria-filled bags.

rangement, insulated with scoria-filled bags partially supported by the poles.

This dome is also covered with a different type of plaster, which is water resistant and insulative: a mixture of paper and cement or lime, or a combination of the two, often called "papercrete" or fibrous cement.

The house is a hybrid, a juxtaposition of different materials, as well as a network of interconnecting structures allowing the area of the house to be quite large without using one single dome.

Kelly Hart describes part of the construction process: "The sloping site was leveled and dug down about a foot below the eventual floor level, which was 5 feet below ground level. Then 6 to 8 inches of loose scoria was put over the entire building area to serve as insulation and good drainage. There is no other foundation or drainage, since the soil is pure, fine sand. We tried using the on-site dry sand to fill our polypropylene bags (misprinted rice bags), but soon discovered that at a certain point after about five feet of stacking, the bags would not hold their shape and without buttressing the wall would collapse. [Author's note: When building with very sandy soils, which consist of round particles of sand, a test structure should be built prior to construction of the main house, with bags not less than 20 inches

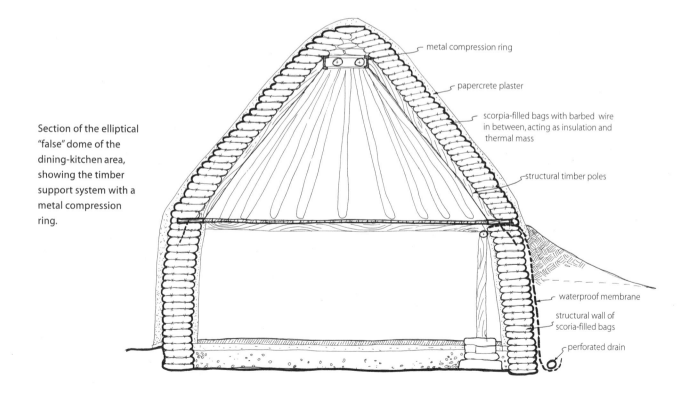

Section of the elliptical "false" dome of the dining-kitchen area, showing the timber support system with a metal compression ring.

metal compression ring

papercrete plaster

scorpia-filled bags with barbed wire in between, acting as insulation and thermal mass

structural timber poles

waterproof membrane

structural wall of scoria-filled bags

perforated drain

(500 millimeters) wide, making sure the sand is damp when filling the bags, well tamped and buttressed.] So we switched to scoria, which is a much lighter material and will pack into a tight, stable wall. Because of the lightness of the scoria, in addition to the usual four-point barbed wire running between each course of bags, we tied the bags to each other with poly baling twine [see photo on page 54], which provides a matrix of fabric across the entire wall to resist fragmenting. It also gives the finish plaster something to grab on to (especially important on interior walls, where in domes you are working against gravity).

"We used individual bags as building blocks, rather than using long tubes of material, because the small bags are easy for one person to carry, and when we tried putting the scoria in long tubes,

With an old woodstove in the foreground, the dining area under the loft can be seen beyond. Flagstones are set in adobe in the foyer. Beyond the table is a windowseat constructed with earthbags.

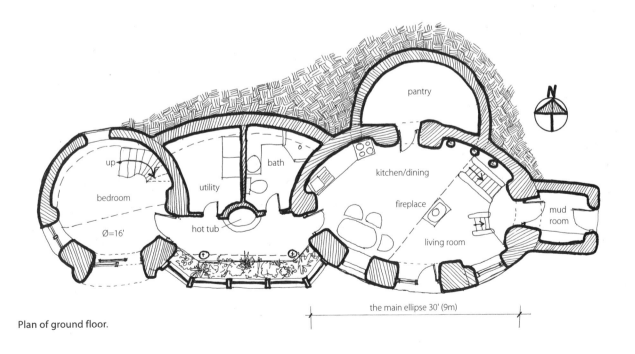

Plan of ground floor.

Kelly applying papercrete to the outside of the dome. This doorway has a span of 6 feet made possible using the double-bag, cross-hatch method of scoria arch.

COURTESY OF HARTWORKS, INC.

Kelly explains, "With the pantry dome open, you can see how the bags are stacked, those long rafters helping to support the bags. Because of the relatively shallow pitch, bermed with earth on the outside, and the considerable weight it would bear, I decided to use logs. Before the pantry was backfilled with sand, I put on two layers of 6-mil plastic sheeting. The dome has not leaked and stays around 40 to 50 degrees Fahrenheit."

COURTESY OF HARTWORKS, INC.

there was a tendency for the whole thing to roll off the wall. [Author's note: This was due to the fact that the tube was narrow and scoria cannot be made compact in the same way that slightly damp earth or sand can. Therefore it may be advisable not to fill the tube completely but loosely, to allow for tamping. Also, always carry the fill to the bag, eliminating the need to lift.] With the individual bags, the bottom seam provides a distinct flat orientation that resists rolling.

"Bags were stacked over a wooden form for the arched doorways. We used a plastic pipe to make air vents. Many of the windows were created using old wagon wheels or culvert couplers to hold the circular shape, and since most of the windows are not operable, the glass used was often unclaimed or cut to the wrong size; these thermal panes can be found cheaply at glass shops. They were imbedded directly in the papercrete (which is so dimensionally stable it will not break the glass), and a second, final coat of papercrete with sand was applied to the outside, overlapping the glass.

"The space connecting the two domes is framed with wood on the south side, while the north side is an insulating scoria-bag wall in the shape of a semicircle. Because of the expanse of glass in this greenhouse section, and the need for a flat place to mount the solar water panels and PV modules, I chose to use conventional wood framing on the south. The north side of the house is bermed with about four feet of earth, and a mound of earth covers the pantry in the north.

"The final layer on top of the papercrete is lime-silica, sand, and white Portland cement plaster.

"Because of the elliptical shape, this dome is not as inherently stable as it would be if it were circular, which is why I used the timber pole. The forces are not uniform, so I had to struggle to keep the shape I wanted. I would not recommend that anyone repeat this experiment, though I do love the shape and feeling as it has turned out."

Kelly has made a video of this construction process (see the resources list).

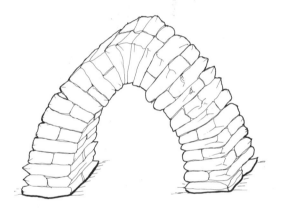

Unbuttressed arch. To create an arch spanning more than 3 feet using lightweight scoria bags, Kelly devised the crosshatched, double-bag arch system shown here.

The Honey House with sculpted gutters and grass seeded in the final layer of earthen plaster.

KAKI HUNTER AND DONI KIFFMEYER'S HONEY HOUSE, UTAH

This house was designed and built in Moab, Utah, by Kaki Hunter and Doni Kiffmeyer of the pioneering firm OK OK OK Productions, who have elevated earthbag construction to a precision craft. The structure is a corbeled earthbag dome with a 12-foot (3.6 meter) interior diameter. Forty tons of earth were needed to fill the bags, and 9 more tons were used for sculpted adobe and for a "living thatch" roof. The gutter system is sculpted adobe, which steers the water around windows,

down buttresses, and away from the foundation.

No cement was used in this structure!

The bags used were small "misprint" bags of two sizes: 17 inches wide by 30 inches long and 22 inches wide by 36 inches long. The fill material used was "reject" sand from a local gravel yard. This had the ideal ratio of 25 percent clay to 75 percent sand for rammed earth construction. The time taken for filling and tamping the bags averaged four of the smaller bags per person per hour, so a team of four averaged sixteen bags per hour, and about ninety-six bags in a 6-hour day. The Honey House is made of eight hundred bags overall. The whole structure took just nineteen days to complete.

The earthen plaster was made with 7 tons of cob, which took seven days to apply all the way up to the roof, which has a 6-inch base for a "live cob thatch roof," where living grass roots were mixed into the final layer of roof plaster.

First floor under construction.

The interior has a sunken floor, 2 feet (60 centimeters) deep. The earthbag wall starts at this level; there is no concrete foundation. Plastic bags were wrapped around the exterior of bags below grade for waterproofing from ground moisture, then backfilled with gravel and earth. The interior walls are earth plastered with local white clay and milk-based *alis* paint. The earthen floor is poured adobe over 4 inches (100 millimeters) of gravel and stone with mud mortar, sealed with a natural, oil-based floor finish.

Except for utilities, windows, and doors, the cost of this house was $1,020, which was the cost of the bags (a quarter of the final cost, including delivery), home-made tools, plywood arch forms (which are reusable), chicken wire, backhoe rental, twenty bales of straw for the earthen plaster, two rolls of barbed wire, and the 40 tons of sand, delivered.

In their manual *Earthbag Construction* (see the bibliography), Kaki and Doni have described their method this way, "We have adopted the FQSS stamp approval—Fun, Quick, Simple, and Solid. By following this criterion, we have made the *ease of the construction process* our priority. As long as the work is Fun and Simple, it goes Quickly and the results are Solid. When the work becomes in any way awkward, FQSS deteriorates into Frustrating, Quarrelsome, Slow, and Stupid, prompting us to stop, change tactics, or blow the whole thing off and have lunch (returning refreshed often spontaneously restores FQSS approval)."

THE LODGE "NJAYA," MALAWI

In structures like the bathing room in the backpackers' lodge "Njaya" in Malawi, in southeast Africa, Adrian Bunting experimented with the sandbag technology prior to constructing a larger sandbag project—an eco-lodge in southern Tanzania. As Adrian explains, "I was trying to get the hang of the sandbag technology based on a couple of photos on the Web. The beach in Tanzania is as remote as you can probably get, so the plan was to build as much as possible with the available materials, basically sand and coconut trees. In Africa, concrete is what everybody wants to build with, but a lot of this is done with lime obtained from live coral reefs, sustaining huge damage. It's also very expensive. I was sent to Malawi to see if the sandbag technology was a viable option for constructing chalets in Tanzania. As far as I could see, it was worth trying if only to achieve the thermal insulation these buildings provide, and also the ease of sealing against mosquitoes, since any timber building is impossible to make bug-free."

The construction procedure used in Malawi is very different from the method described in this book. As Adrian explains, "The bags are filled with pure dry lake sand, of a granite type. There is no barbed wire, and I did find the bags slipped during construction, so the roof has reinforcement bars bent and used as a permanent form. The corridor is just stacked in an arch.

"You might be interested in a conversation I had with the builder when we started the roof. He was shaking his head, so I asked him why. He replied that this wasn't a roof. I asked him why, and he replied, 'A roof is made of tin.' I said, but tin is noisy when it rains, it heats up quickly in the sun, and when it's old mosquitoes get in. 'It is still a roof,' he replied. But this is ten times cheaper, I pointed out. 'I see,' he said. . . ."

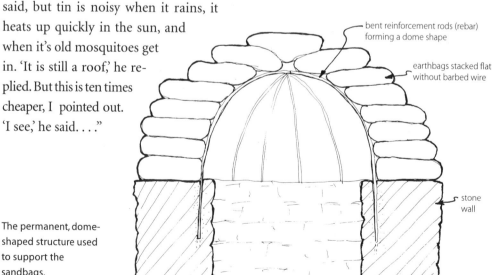

bent reinforcement rods (rebar) forming a dome shape

earthbags stacked flat without barbed wire

stone wall

The permanent, dome-shaped structure used to support the sandbags.

The New House of the Yaquis, Mexico

The indigenous Yaqui community of Pueblo de Sarmiento is located on the outskirts of the city of Hermosillo, capital of the Mexican state of Sonora, which borders the United States. Mexico's government gave a ten-acre parcel of undeveloped land to the Yaquis of Hermosillo in 1995. This land is a desert at the bottom of the surrounding hills.

Giovani Panza, of the organization Itom Yoemia Vetchivo, whose mission is to find funding for projects that improve the Yaqui community's living conditions, became acquainted with Pueblo de Sarmiento through a conference of indigenous peoples of the Mexico–U.S. border, which was held in Hermosillo. In the spring of 1997, Itom Yoemia Vetchivo received a grant of four thousand dollars to build the Yaqui community a prototype low-cost shelter, and they approached Cal-Earth for help. With two volunteers experienced in conventional construction who were carrying out apprenticeships at Cal-Earth, I offered to coordinate this project, which was organized as a three-week volunteer project.

The people of Pueblo de Sarmiento usually build their own homes out of whatever is on hand: cardboard, corrugated asphalt sheeting, and small pieces of found timber. These structures need to be rebuilt after the seasonal hard rains. They had one source of fresh water coming into the village, and no sanitation facilities apart from an outhouse. In spite of the low standard of living, we remarked upon a sense of cheerfulness all around.

Taking stock of the materials available on-site, we saw that one of the few resources that was abundant was earth—very sandy with hardly any clay—so the earthbag technology would be very appropriate.

The house was to be a Nader Khalili-designed, low-cost prototype, the "three-vault house," which utilizes a simple design based on the repetition of single arched units, simplifying construction. Khalili's arrangement of vaults eliminates the need for corridors, and additional vaults can be added later easily. Residents have a view through the depth of two vaults at one time, increasing the sense of interior spaciousness, and variety can be introduced through placement of windows and other elements such as niches and alcoves. A wind catcher faces prevailing summer breezes to direct air into the house for cooling. In addition, the vaulted curve of the roof creates shaded zones of cooler air, while the sun's path overhead encourages air movement inside house by gradually

Top: A house constructed of corrugated asphalt, which becomes unbearably hot in the Mexican sun.

Above: Prototype of the three-vault house.

shifting the shaded zone up and over the vault.

Contrary to my intentions, the people of Sarmiento were asked to prepare a concrete foundation prior to our arrival, which was the first of many unnessary measures in this project. Due to this extra work, time was lost.

A team of fifteen Yaqui workers had assembled for the project. Some were from this community, and others came from the village of Ciudad Obregon, farther south. They were all to receive standard Mexican wages, which to us meant that their motive for working hard would not be primarily educational. Yet with all this available labor, I could not have foreseen what could go wrong.

Once the concrete of the footings had cured, the work began. At first, earthbag construction seemed very strange to the Yaquis, and reluctantly, but with great merriment, they embarked on this adventure with space-age technology. Their first day of laying bags was very slow, but soon they gained experience and speed. The Cal-Earth advisers, who were not themselves accustomed to alternative construction, had insisted on putting 12 percent cement into the earthen fill, thereby treating the bags as concrete forms instead of rammed earth. In six days, the laying of the earthbags was complete and the structure was ready for the vault construction. Over the whole week, the speed of the bag laying averaged 23 feet per hour per team, with three people in each team.

At this stage, we faced a huge dilemma—whether to build some kind of timber formwork (the cost of which would come to more than what was budgeted for the whole house), or instead to try and devise a way of constructing the vault using permanent, built-in formwork. "What

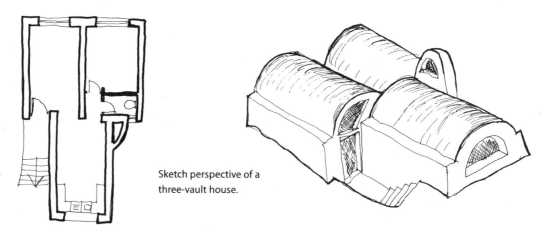

Sketch plan of a three-vault house.

Sketch perspective of a three-vault house.

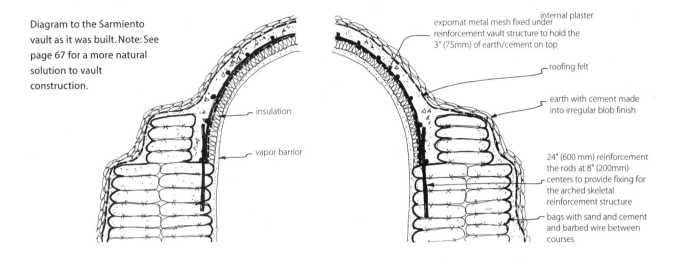

Diagram to the Sarmiento vault as it was built. Note: See page 67 for a more natural solution to vault construction.

insulation

vapor barrier

internal plaster

expomat metal mesh fixed under reinforcement vault structure to hold the 3" (75mm) of earth/cement on top

roofing felt

earth with cement made into irregular blob finish

24" (600 mm) reinforcement the rods at 8" (200mm) centers to provide fixing for the arched skeletal reinforcement structure

bags with sand and cement and barbed wire between courses

is the point of building a house without any wood if you are going to use vast amounts of wood for the formwork?" was the commonsense question. A way of building a vault using reinforcement rods, expanded metal mesh, and rigid foam insulation was devised. "What happened to the idea of natural, alternative construction?" was my constant question.

Unfortunately, the concept of empowerment through use of the simplest techniques and the most available materials was almost lost, because the local people were unable to build more of these buildings due to the vast cost of the materials used in the prototype. So what went wrong? Why did the project's cost escalate from the four thousand budgeted to ten thousand dollars? The most expensive material costs of the project were: *cement*, for foundations, walls, and roof; *metal*, for

the roof structure; and *rigid foam*, for the insulation. There were numerous possible alternatives. For instance, the foundations could have been inexpensive gravel- or rubble-filled trenches; or two stabilized earthbag rows laid directly on undisturbed ground, since this is not an earthquake area; or two courses of bags filled with gravel laid on undisturbed ground (see the descriptions of earthbag foundations in chapter 3).

There was no need to put any cement in the fill for the walls. The earthbag technology has been specifically developed for building in areas with no clay or wood, and using unstabilized earth would have been entirely practical at Sarmiento. For the roof structure, it would have been better to use the locally grown, bamboo-like carrizo reed for vaults (see page 67), or corrugated metal sheeting, which is cheap and easily

Top, and top, right: When the reinforcement rods are fixed in their arched shape, the structure is strong and resilient.

Above: Expomat mesh fixed under the reinforcement rods to form a rough surface for the soil-cement coating layered on top.

Right: View from vault 1 of the proposed entrance area and the intersection of vaults 1 and 2.

available. As for insulation, a straw-clay mix would have been preferable to expensive rigid foam (see the discussion of roofs in chapter 5).

This project underscored the truth that a thorough understanding of any building technique is necessary to avoid unnecessary cost and complexity. Moreover, when working with builders who speak different languages, it is important to acquire the basics of the local people's language in order to work effectively. Unless your translator has extensive knowledge of the building process, it will be difficult to have any in-depth communication.

Yet in spite of our feeling that we made several big mistakes, the project in many ways was a success. The Yaqui workers from Hermosillo had never worked so closely with Yaquis from other areas, nor with foreigners. The process was truly a communal experience, marked by continuous problem-solving and endless laughter. Everybody was learning one or two of the languages that were being spoken at all times. Yaquis were actively involved in the formulation of the budget and the purchase of the materials(Kari 1997). Ultimately, the "New House of the Yaquis" was truly a surprise to everybody. It was built by the concentrated labor of the whole community, and because the effort was demanding of everyone, it left a deep impression on every person involved.

AFTERWORD

Two of the new owners of the three vault house.

To live in a natural house is a privilege.

Through the process of natural building, we can reconnect with the basics. We can find simple solutions to seemingly complex problems, often allowing nature instead of machinery to do the work. We can create communities that reconnect us with the earth and with each other, for earth itself is the most wonderful material, feeding us and housing us. And to possess knowledge of some of earth's mysteries is a great gift, allowing enormous freedom.

I hope that all those looking through the pages of this book draw inspiration to create their own home, and to adapt their house to its unique, individual, and distinctive environment.

It's your game, it's your joy—go and play!

BIBLIOGRAPHY

Andreson, Frank. "Oh Muddy Clay, O Clayish Loam: Introduction to German Clay Building Techniques." *Joiners Quarterly: The Journal of Timber Framing and Traditional Building* 37 (1997): 18-23.

At Home with Mother Earth. Video recording. Los Angeles: Feat of Clay, 1995.

Beale, Kevin. *Factsheet on Straw Bale Construction.* Machynlleth, Wales: Centre for Alternative Technology, 1997.

Bedford, Paul, Bruce Induni, Liz Induni, and Larry Keefe. *Appropriate Plasters, Renders and Finishes for Cob and Random Stone Walls in Devon.* Exeter, U.K.: Devon Earth Building Association, 1993.

Bee, Becky. *The Cob Builders Handbook: You Can Hand-Sculpt Your Own Home.* Cottage Grove, Oreg.: Groundworks Press, 1997. Distributed by Chelsea Green.

Billatos, S. B., and N. A. Basaly. *Green Technology and Design for the Environment.* London: Taylor & Francis, 1997.

CRATerre, with Hugo Houben and Hubert Guillaud. *Earth Construction: A Comprehensive Guide.* Villefontaine Cedex, France: Intermediate Technology Centre, 1994.

Crews, Carol. *The Art of Natural Building: Earth Plasters and Aliz.* Kingston, N.M.: Network Productions, 1999.

Day, Christopher. "Human Structure and Geometry," *Eco-Design Journal* 3, no. 3 (1995).

_____. *Places of the Soul: Architecture as a Healing Art.* Glasgow: Collins, 1990.

Denyer, Susan. *African Traditional Architecture: A Historical and Geographical Perspective.* New York: Africana, 1978.

Devon Historic Building Trust. *The Cob Buildings of Devon 1: History, Building Methods, and Conservation.* Exeter, UK: Devon Historic Building Trust, 1992.

Droege, Sam. "'How to' book, 1861," archived in 1997 on CREST's Straw-bale Listerv: <http://solstice.crest.org/efficiency/strawbale-list-archive/index.html>.

Easton, David. *The Rammed Earth House.* White River Junction, Vt.: Chelsea Green, 1996.

Escott, Carol, and Steve Kemble. "Earthbag Construction in the Bahamas." *Earth Quarterly* 2 (1997): 24-27 .

Farrelly, David. *The Book of Bamboo.* London: Thames & Hudson, 1996.

Fathy, Hassan. *Architecture for the Poor: An Experiment in Rural Egypt.* Chicago: University of Chicago Press, 1986.

Frei, Otto. *Tensile Structures.* Cambridge: M.I.T. Press, 1969.

Hahn, Tom. "Good Shoes: Foundations." *The Last Straw* 16 (1996):1-15.

Hart, Kelly, and Rosana Hart. Video recording. *A Sampler of Alternative Homes: Approaching Sustainable Architecture.* Crestone, Colo.: Hartworks, 1998.

Hidalgo, Oscar L. *Manual de Construccion con Bambu: Construction rural - 1*. Bogota: Universidad Nacional de Colombia, 1981. [Estudios Techicos Colombianos Ltda., Apartado Aereo 50085, Bogota, Colombia.]

Hunter, Kaki and Doni Kiffmeyer. *Earthbag Construction*. Moab, Utah: OK OK OK, Productions, 2000.

Kachadorian, James. *The Passive Solar House: Using Solar Design to Heat and Cool Your Home*. White River Junction, Vt.: Chelsea Green, 1997.

Kanuka-Fuchs, Feinhard, and Jennifer Ratenbury. *Biological Natural Organic Paints and Surface Treatments*. Auckland, N.Z.: Building Biology and Ecology Institute of New Zealand, 1991.

Keefe, Larry. *The Cob Buildings of Devon 2: Repair and Maintenance*. Exeter, U.K.: Devon Historic Building Trust, 1993.

Kemble, Steve. *How to Build Your Elegant Home with Straw Bales: A Guide for the Owner-Builder*. Bisbee, Arizona.: Sustainable Systems Support, 1995.

Kennedy, Joseph F. *The Art of Natural Building: Design, Construction, Technology — Compiled from Presentations at The Natural Building Colloquia Southwest*. Kingston, N.M.: Network Productions, 1999.

_____. "Expanding the Horizons of Earthbuilding: New Research from the California Institute of Earth Art and Architecture." *Adobe Journal* 11 (1994): 16-21.

Khalili, Nader. *Ceramic Houses and Earth Architecture: How to Build Your Own*. London: Harper & Row, 1986; Los Angeles: Burning Gate Press, 1990; Hesperia, Calif.: Cal-Earth Press, 1996.

_____. "Lunar Structures Generated and Shielded with On-Site Materials." *Journal of Aerospace Engineering* 2, no. 3 (1989).

King, Bruce. *Building of Earth and Straw: Structural Design for Rammed Earth and Straw Bale Architecture*. Sausalito, Calif.: Ecological Design Press, 1996. Distributed by Chelsea Green.

Minke, Gernot. *Earth Construction Handbook: The Building Material Earth in Modern Architecture*. Southhampton and Boston: WIT Press, 2000.

Mollison, Bill. *Permaculture: A Designer's Manual*. Tyalgum, Australia: Tagari, 1988.

Muller, R. "Hesperia's Domes Pass Seismic Tests." *The Daily Press*, 30 September 1993.

Myhrman, Matts, and S. O. MacDonald. *Build it with Bales: A Step-by-Step Guide to Straw-Bale Construction — Version Two*. Tucson: Out on Bale, 1999. Distributed by Chelsea Green.

Nerburn, Ken. *Neither Wolf nor Dog: On Forgotten Roads with an Indian Elder*. San Rafael, Calif.: New World Library, 1994.

Norton, John. *Building with Earth: A Handbook — Second Edition*. London: Intermediate Technology, 1996.

Oliver, Paul. *Encyclopedia of Vernacular Architecture of the World*. Cambridge: Cambridge University Press, 1997.

Outram, Iliona. "The Best of Times for Earth Builders: Research at Cal-Earth." *Adobe Journal* 12/13 (1996): 52-58.

_____. "Interview with Nader Khalili." *Adobe Journal* 12/13 (1997): 54.

Panza, Giovani, et al. *U Vemela Hiakim Kari, The New House of the Yaquis*. Tucson: Itom Yoemia Vetchio, 1997.

Pearson, David. *The Natural House Catalog*. New York.: Simon Schuster, 1996.

Potts, Michael. *The New Independent Home: People and Houses that Harvest the Sun, Wind, and Water*. White River Junction, Vt.: Chelsea Green, 1999.

Rapoport, Amos. *House, Form and Culture.* Englewood, N.J.: Prentice Hall, 1969.

Richardson, T. L., and E. Lokensgard. *Industrial Plastics: Theory and Application.* Albany: Delmar, 1989.

Rigassi, V. *Compressed Earth Blocks, Volume 1: Manual of Production.* Eschborn, Germany: GATE, 1995.

Rudofsky, Bernard. *Architecture without Architects: A Short Introduction to Nonpedigree Architecture.* Albuquerque: University of New Mexico, 1964.

Schofield, Jane. *Lime in Building: A Practical Guide.* Crediton, U.K.: Black Dog Press, 1994.

Smith, Michael G. *The Cobber's Companion: How to Build Your Own Earthen Home.* Cottage Grove, Oreg.: The Cob Cottage Company, 1998.

Solberg, Gordon. "A Colorado Papercrete Tour." *Earth Quarterly* 4 (1999): 14-21.

_____. "Fibrous Cement: A Revolutionary Building Material." *Earth Quarterly* 1 (1997): 2-15.

_____. "How To Build a Papercrete Mixer." *Earth Quarterly* 2 (1998): 12-18.

Spence, R. J. S., and D. J. Cook. *Building Materials in Developing Countries.* Chichester and New York: John Wiley and Sons, 1983.

Steen, Athena, and Bill Steen. *The Beauty of Straw Bale Homes.* White River Junction, Vt.: Chelsea Green, 2000.

_____. *Earthen Floors.* Elgin, Ariz.: Canelo Project, 1997.

_____. "The Straw Bale Earthen House." *Adobe Journal* 12/13 (1997): 42-45.

Steen, Athena, Bill Steen, and David Bainbridge, with David Eisenberg. *The Straw Bale House.* White River Junction, Vt.: Chelsea Green, 1994.

Stern, Ephraim. *Dor: Ruler of The Seas.* Jerusalem: Israel Exploration Society, 1994.

The Straw Bale Solution. Video recording. Santa Cruz, N.M.: Networks Production, 1998.

Stulz Roland, and Kiran Mukerji. *Appropriate Building Materials: A Catalogue of Potential Solutions.* St. Gallen, Switzerland: SKAT Publications, and London: IT Publications,1993.

Swan, James, and Roberta Swan. *Dialogues with the Living Earth: New Ideas on the Spirit of Place from Designers, Architects, and Innovators.* Wheaton, Ill.: Quest Books, 1996.

Thomas, Lewis. *The Lives of a Cell.* New York: Viking Press, 1974.

Thompson, Kim., et al. *Straw Bale Construction: A Manual for Maritime Regions.* Ship Harbour, Nova Scotia, Canada: Beautiful Sustainable Buildings, 1995.

Tibbets, Joseph M. *The Earthbuilders Encyclopedia.* Bosque, N.M.: Southwest Solaradobe School, 1989.

Tibbles, R. "Building with Hemp." *Building for a Future* (1997/98): 16.

Trimby, Paul. *Solar Water Heating: A DIY Guide.* Machynlleth, Wales: Centre for Alternative Technology, 1996.

Van der Ryn, Sim. *The Toilet Papers: Recycling Waste and Conserving Water.* Sausalito, Calif.: Ecological Design Press, 1994. Distributed by Chelsea Green.

Vittore, Phillip. "Dome and Vault Engineering." *Adobe Journal* 12/13 (1997): 56.

Woodward, J. *Nature's Little Builders.* Thailand: Sirivanata Palace Press / Electric Paper, 1995.

Zelov, Chris, and Brian Danitz. *Ecological Design: Inventing the Future.* Video recording. Ecological Design Project, 1996. Distributed by Chelsea Green.

Ziesemann, Gerd, Martin Krampfer, and Heinz Knieriemen. *Naturliche Farben: Anstriche und Verputze selber herstellen.* Aarau, Switzerland: AT Verlag, 1996.

RESOURCES

PERIODICALS

Adobe Builder. Southwest Solaradobe School, PO Box 153, Bosque NM 87006 USA.. Telephone: +(505) 861-1255. Internet: www.adobebuilder.com .

Adobe Journal. Published by Michael Moquin, PO Box 7725, Albuquerque NM. 87194 USA. Telephone: +(505) 243-7801.

Building for a Future. The Association for Environment-Conscious Builders. Nant-y Garreg, Saron, Llandysul, Carmarthenshire, SA 44 5EJ England. Telephone: +01559 370908.

Building with Nature. PO Box 4417, Santa Rosa CA 95402. USA. Telephone: +(707) 579-2201.

Designer/Builder. 2405 Maclovia Lane, Santa Fe NM 87505 USA.. Telephone: +(505) 471-4549.

Earth Quarterly. Box. 23, Radium Springs NM 88054 USA.. Telephone: +(505) 526-1853.

Eco Building Times. Northwest Eco Building Guild, 217 Ninth Ave., North Seattle WA 98109 USA.

Eco Design. PO Box 3981, Main Post Office, Vancouver BC V6B 3Z4 Canada.

Environmental Building News. 122 Birge Street, Suite 30, Brattleboro VT 05301 USA.. Telephone: +(802) 257-7300, internet: www.ebuild.com .

Erosion Control: The Journal for Erosion and Sediment Control Professionals. Published monthly by Forester Communications, Inc., 5638 Hollister # 301, Santa Barbara CA 93117 USA. Telephone: +(805) 681-1300.

Green Building Digest. Queens University of Belfast, 2-4 Lennoxvale, Belfast BT9 5BY Northern Ireland. Telephone: +01232 335466.

Green Connections. PO Box 793, Castlemaine 3450 Australia. Telephone: +(03) 5470 5040.

Home Power: The Hands-on Journal of Home-Made Power. PO Box 14230, Scottsdale AZ 85267-4230 USA. Telephone: +(919) 475-0830.

Joiners Quarterly: The Journal of Timber Framing and Traditional Building. PO Box 249, Brownfield ME 04010 USA. Telephone: +(207) 935-3720.

Permaculture. Permanent Publications, The Sustainability Centre, East Meon, Hampshire GU32 1HR, England. Telephone: 01730 823311, internet: www.permaculture.co.uk .

Positive News. The Six Bells, Bishops Castle, Shropshire SY9 5AA. England. Telephone: + 01588 630 121/122.

The Last Straw: The Grassroots Journal of Straw Bale and Natural Building. HC 66, Box 119, Hillsboro NM. 88042 USA. Telephone: +(505) 895-5400. e-mail: thelaststraw@strawhomes.com, internet: www.strawhomess.com .

The Permaculture Activist. PO Box 1209, Black Mountain NC 28711 USA Telephone: + (828) 298-2812.

ORGANIZATIONS AND COMPANIES IN THE U.S. AND THE U.K.

Frank Andresen
Construction with light clays and clay plasters; also offers dry clay products as well as workshops and consulting.
Kiefernstrasse 2, 4000 Dusseldorf, Germany. Telephone: +0211 7333216.

Kevin Beale.
Design, consultation, and construction using earthbag, straw bale, and other methods.
Ty-Capel-Graig, Talsarnau, Gwynedd, Wales. LL47 6UG, England. Telephone: +(0) 1766 770 696, e-mail: Kevinbeale@breathemail.net.

Black Range Lodge
Videos, educational materials, and resources for straw bale, cob, and other alternative building techniques. Bed & breakfast lodging for educational retreats.
Star Route 2, Box 119, Kingston NM 88042 USA. Telephone: +(505) 895-5652, internet: www.epsea.org/straw.html

Building Biology and Ecology Institute of New Zealand
22 Customs Street West, PO Box 2764 CPO, Auckland, New Zealand. Telephone: +(64-9) 358 2202.

California Institute of Earth Art and Architecture (*Cal Earth*)
Founded by Iranian Architect Nader Khalili to pursue research in sustainable human shelter principally through earthen materials and an earthbag technique called "Superadobe" for domes and vaulted structures. Apprenticeship retreats, and weekend visitations to the demonstration site.
Cal-Earth, 10225 Baldy Lane, Hesperia CA. 92345. USA. Telephone: +(1) 760 244 0614, e-mail: CalEarth@aol.com, internet: www.Calearth.org .

Canelo Project
Set up by the co-authors of The Straw Bale House. *Offering comprehensive straw bale construction workshops and educational resources, with a focus on traditional materials and practices including earthen plasters, floors, and bread ovens. Workshops in southern Arizona and Mexico.*
HC1, Box 324, Elgin, Arizona 85611 USA. Telephone: +(520) 455-5548, e-mail: absteen@dakotacom.net, internet: www.caneloproject.com

Centre for Alternative Technology
Workshops and educational resources, with a large bookshop for alternative construction and sustainable living information.
Machynlleth, Powys SY20 9AZ, Wales, UK. Telephone: +01654 702400, e-mail: cat@gn.apc.org, internet http://www.cat.org.uk

The Cob Cottage Company

Workshops and resources in cob construction and passive solar design.
PO Box 123, Cottage Grove, OR 97424 USA. Telepone: +(541) 942-2005, intenet:
www.deatech.com/cobcottage

Construction Resources

Specializing in ecological construction techniques. Exhibition center, resources, lectures.
16 Great Guildford Street, London SE1 OHS UK. Telephone: +020 7450 2211.

Constructive Individuals

Architects specialising in alternative construction, self-build projects, and construction workshops.
London, UK. Telephone: +020 7515 9299.
CRATerre–EAG School of Earth Construction
Maison Leurat, Rue du Lac, BP 53, F-38092, Villefontaine Cedex, France.

CRG Design Healthy Homes

Supplier of natural building materials. Design and consultation services.
Cedar Rose, PO Box 113, Carbondale CO 81623 USA. Telephone: +(970) 963-0437,
e-mail: crose@rof.net .

Development Centre for Appropriate Technology

Building code information and educational resources.
David Eisenberg, PO Box 27513, Tucson AZ 85726 USA. Telephone: +(520) 624-6628, e-mail:
info@dcat.net, internet: www.dcat.net .

Earth Hands & Houses, and PWA Architects

*Design, consultation, workshops and construction of sustainable, ecological, 'organic' projects in
developed and developing countries.*
Paulina Wojciechowska, Architect. 18 The Willows, Byfleet, Surrey KT 14 7QY England.
Telephone: + (0) 1932 352129, e-mail: EHaH@excite.co.uk, internet:
www.EarthHandsAndHouses.org .

Earthwood Building School

*Resources, workshops, and design consultations for cordwood-masonry construction, stone circles,
mortgage-free living, and off-the-grid energy strategies.*
Rob and Jaki Roy, 366 Murtagh Hill Road, West Chazy NY 12992 USA. Telephone: +(518) 493-
7744.

Gourmet Adobe

Specializing in clay slips with mica and adobe.
Carole Crews, HC 78, Box 9811, Ranchos de Taos, NM 87557 USA

Hartworks, Inc.

*Producers of a two-hour video (see the Bibliography) available in U.S. standard NTSC VHS
format. Includes a section on earthbags, as well as covering other natural building techniques.
Another video specifically on earthbag construction is in production.*
Kelly and Rosana Hart, Hartworks, Inc., PO Box 632, Crestone, CO 81131, USA. Telephone:
+(719) 256 4278, e-mail: Office@hartworks.com, internet: www.hartworks.com .

Heartwood School

Johnson Hill Road, Washington MA 01235 USA. Telephone: +(413) 623-6677.
Willbheart@aol.com

Imagination Works

Dominic Howes, builder and consultant of alternative home construction using earthbags and other alternative methods.
P.O. Box 477, Dragoon, AZ 85609 USA.. E-mail: dominichowes@mailcity.com, internet: www.sfnet.net/imagination .

Intermediate Technology Centre

Bookshop and educational resources.
103-105 Southampton Row, London WC1B 4HH, UK. Telephone +020 7436 2013.

International Institute for Bau-Biologie & Ecology.

PO Box 387, Clearrwater, FL 33757 USA. Telephone: +(813) 461-4371, e-mail: baubiologie@earthlink.net, internet: www.bau-biologieusaa.com

Joseph Kennedy

Architectural designer, writer, and peripatetic scholar of natural building and ecological design. Teaches, gives workshops and consultations.
Star Route 2, Box 119, Kingston, NM 88042 USA. Telephone: +(505) 895 5652, e-mail: livingearth62@hotmail.com .

OK OK OK Productions

Providing earthbag construction, training, along with workshops on wild clay and lime plasters, earthen floors. Design consultations for dome and arch construction.
Kaki Hunter & Doni Kiffmeyer, 256 East 100 South, Moab UT 84532 USA. Telephone: +(435) 259-8378, e-mail: okokok@lasal.net .

Out on Bale by Mail (un)Ltd.

Straw bale consultation, educational programs, wall raising supervision, and bulk orders of the book Build It With Bales *(see the bibliography).*
2509 N. Campbell, #292, Tucson, AZ 85719 USA.

Sustainable Systems Support

Design, consultation, workshops. Specializing in earthbag and straw bale construction methods. Source of printed and video resources.
Carol Escott and Steve Kemble, PO Box 318, Bisbee, AZ 85603 USA. Telephone: +(520) 432-4292, e-mail: sssalive@primenet.com, internet: www.bisbeenet.com/buildnatural/ .

Women Build Houses

Workshops, referrals, and tool library.
1050 S. Verdugo, Tucson AZ 85745 USA. Telephone: +(520) 882-0985, e-mail: wbhwbhwbh@aol.com .

EQUIPMENT AND SUPPLIES

Continuous berm machine for filling earthbags

Can extrude a continuous berm at rates of 10 to 50 feet per minute. "No trenching or stacking required. With weight typically exceeding 100 pounds per foot, the continuous berm conforms tightly to underlying soil surfaces, will not blow over, and is extremely difficult to dislodge from original placement location. Additionally the berm can be cut into sections and stacked for streambank stabilization, 'sand bagging,' fluids containment, or used separately for check structures." *(description from* Erosion Control *journal.)*

Available from Innovative Technologies, PO Box 378, 250 Hartfford Road, Slinger, WI 53086-0378 USA. Telephone: +(414) 644-5234.

Auro Products

For casein, natural paints, oil solvents, waxes, and other finishes.

Sian Company, PO Box 857, Davis, CA 95617-3104 USA. Telephone: +(916) 753-3104, internet: www.dcn.davis.ca.us/go/sinan/auroinfo.html or www.auro.de/

Livos Phytochemistry of America

Natural, nontoxic paints and stains.

13 Steeple Street. PO Box 1740, Mashpee MA 02649, USA.

Tel: 508 477 7955. www.livos.com for natural paints and wood finishes.

Livos UK

Unit 7 Maws Croft Centre, Jackfield, Ironbridge, TF8 7LS UK. Telephone: +0 1952 883288

INDEX

A

acidic materials, 81, 83, 92, 106
additions, building, 17, 32–33
additives. *See* stabilizers
adobe, 4–8, 140
    bitumin, use of, 89
    cracks allowed, 105
    earthbag foundation, 38
    finishes, 76, 80, 90
    floors, 118–19
    furniture, 114
    roofs, 66
    soil mix, 17, 46, 103, 107
Africa, 6, 16, 18, 19, 22, 142
Ahlquist, Allegra, 56, 127
alis, 95–97, 114, 141
alkaline materials, 81, 83, 84, 106
animal product stabilizers, 79, 82, 106
apses, 22, 26
arched openings, 13, 14, 21, 22, 58–60
arches, xvi, 13, 21–24, 138–39. *See also*
    vaults
    for additions, 17
    buttresses for, 23–24, 57, 60
    corbeled, 26
    keystone, 23, 24, 59–60
    lancet (catenary), 22–23, 44
    unbuttressed, 139
Australia, 10

B

bags. *See* earthbag bags
Bahamas, 16, 131–34
banana leaf juice, 81
barbed wire, 36, 46, 54, 58, 137

beeswax, 119–20
benches. *See* furniture
bentonite clay, 88–89
binders. *See also* clay
    casein as, 98–99, 100
    in limewash, 97
    in paint, 95
    as sealant, 93
    soil mix using, 103, 105–109
bitumen, 79, 89
bond beams, 15, 18, 46, 60–63, 66, 79,
    123, 127, 130–34
Building Biology and Ecology Institute
    of New Zealand, 98
building codes, 4, 14, 90, 105
Bunting, Adrian, 142
burlap bags
    for earthbags, 18, 43–45, 76
    for insulation, 110
buttresses. *See also* tension rings
    for arches, 23–24, 57, 60
    domes, 26, 44, 54–57, 126, 136
    dry area, construction in, 35
    sand-filled earthbags, 49, 136
    spring line, 23, 26
    straight walls, 127

C

California, 14–15, 24, 34, 105
California Institute of Earth Art and
    Architecture (Cal-Earth), xiv–
    xviii, 14, 43, 124
Canelo Project, Mexico, 71–72, 100, 116
casein, 82, 97–100
catenary (lancet) arch, 22–23, 44

cement
    bond beams, 46, 79
    compression rings, 46, 79
    environmental impact, 46, 79, 89, 92
    finish stabilizer, 46, 66, 75, 78–79,
        89–91, 100–101, 136
    stucco, 90, 123, 127
cement-stabilized earthbags, 37, 38, 46,
    62, 89, 123, 144–45
Center for Appropriate Technology, 15
Ceramic Houses (Khalili), 21–22, 26
clay-based building materials, 5–6, 19,
    46. *See also* adobe; cob construc-
    tion; straw-clay blends
    as binder, 6, 103, 105–109
    bitumen used with, 89
    interior walls, floors, furniture, 103
    moisture control and, 76
    soil analysis, 80, 103–105
clay finishes
    erosion of, 77
    interior plaster, 92
    as sealant, 93
    waterproofing, use for, 88–89
clay in earthbags, 13, 17, 18, 49
    cement stabilization of, 89
    damp areas, construction in, 35–38,
        49
    lime stabilizer use and, 83
    polypropylene earthbags, use of, 75
clay slip, 95–97, 114, 141
Cobber's Companion, The (Smith), 70,
    104
cob construction, 4, 5, 6, 8–9, 141
    earthbag foundation, 16

finishes, 76, 83, 90, 97
furniture, 114, 115
soil mix, 17, 46, 103, 107–108
thatched roofs, 68–70
Cob Cottage Company, 109
cold climates
building orientation for, 31–32
earthen plasters, effect on, 77
finish stabilizers, use of, 79
hybrid house for, 128–29
insulation for, 111
lime plasters, application of, 87
soil mix for, 106
compass, construction, 34, 48–49
compression rings, 26, 46, 54, 57, 79, 123, 125, 136
concrete. *See also* cement
bond beams, 18, 60–63, 123, 127, 130–34
foundations, 38, 127, 144–45
connectors. *See* passageways
Constructive Individuals, xiv
corbeling, 16, 26–27, 43–44, 49, 110, 135
cordwood masonry, earthbag foundation for, 38
costs, 13
Ahlquist square house, Arizona, 127
of earthbag bags, 45
Escott-Kemble house, the Bahamas, 134
Hunter-Kiffmeyer house, 141
Njaya backpackers' lodge, Malawi, 142
Tassencourt dome, Arizona, 125
Yaquis vault house, Mexico, 145, 147
CRATerre, 10, 78–79, 81
Crews, Carol, 82, 90, 95–96, 107

D
damp areas, construction in, 13, 35–40, 49, 66. *See also* flood-prone areas, building in; waterproofing; water resistance
finishes for, 75

finish stabilizers, use of, 79
roofs for, 65, 71, 80
design considerations, 13, 17, 29–33, 80
Devon Earth Builders, 90, 106–107
Devon Earth Building Association, 97
Devon Historic Building Trust, 108
disaster-relief housing, 13, 16, 43
domes, xvii, 13, 14, 15
brick, 18
buttressing, 26, 44, 54–57, 136
compass for, 34, 48–49
corbeled, 27, 49, 110, 135, 140–41
defined, 22, 26
earthbags, construction using (*See* earthbag construction methods)
earthquake resistance of, 15
examples, 123–26, 135–42
form, use of, 49
openings (*See* openings)
for roofs, 66
shell, determining thickness of, 27
spanned, 22
on square structures, 27
stability, increasing, 49
thickness of shell, determining, 27
doors. *See* openings
drainage, 30, 37, 38, 75, 136
earthen floor, 117
gabions, 30, 39–40
gutters, 140
dry areas, construction in, 35–37, 66
dryboard, 109–10
dry-stone foundation, 40

E
earth architecture, xiv, 3–19. *See also specific building methods, e.g.* adobe
clay-based building materials, 5–6
history of, 3, 5, 6, 22
revival of, 4–5
earthbag bags
cost, 45
making, 45

materials for, 13, 38, 43–45
recycled bags, 44, 45
"seconds," 44, 45, 133, 140
small bags, 52, 137
sources of, 19, 44
earthbag buildings, 13–19. *See also* damp areas, construction in; domes; waterproofing
additions, planning for, 17, 32–33
connecting, 13, 24–25, 136, 139
design considerations, 13, 17, 29–33, 80
in dry area, 35–37, 66
examples, 123–47
finishes (*See* finishes)
labor intensity of, 13
layout of, 13–14
recycling, 17
seismic and structural testing, 14–15
shape of, 32–33 (*See also* domes)
utilities, 32, 120–21, 139
Earthbag Construction (Hunter and Kiffmeyer), 141
earthbag construction methods, xiv, 4, 13, 43–63. *See also* bond beams; buttresses; earthbag fill; foundations; hybrid construction; openings
appropriate use of earthbags, 16–17
cutting into bag, 17
forms, use of, 25, 49, 58–59, 144–46
for furniture, 18, 114
keying, 36, 54, 137
materials, 44–46
simplicity of, 19
structural walls of conventional house, 128–29
tamping, 52–54
tools, 13, 46–49
tying courses together, 49, 137
earthbag fill, 16–17, 49–50, 131, 140
cement-stabilized (*See* cement-stabilized earthbags)
filling process, 45–46, 50–52, 137–39
pH of, 81

earthbag fill, *continued*
  procedure, 140
  soil analysis, 80, 103
  width of filled bag, 45
Earth Building Association, 88
Earth Construction (CRATerre), 78–79,
  81
earthen floors, 103, 107, 115–20
  alternatives to, 118–19, 127
  construction, 118–19
  heating in, 127
  insulation in, 111, 116, 117
  layers, 116–17
  maintenance, 119–20
  repair, 120
  sealants, 93, 117, 119–20
  for upper story, 118
Earthen Floors (Steen), 100, 119
earthen plasters, 66, 75–77, 80
  advantages/disadvantages, 76–77
  application of, 77–78, 91–92
  clay slip *(alis)* finish, 95–97, 114, 141
  for furniture, 114
  interior finish use, 92
  maintenance, 77, 91, 100–101, 109
  nonstabilized, cement plaster on, 90
  permeability, 76, 80, 90, 93
  sealants, 80, 92–94
  stabilizers (*See* stabilizers)
  waterproofing, 66, 88–89
earth-filled tires, 39, 111
Earthmother Dwelling Retreat, xvi–xviii
earthquake resistance, 15, 38, 54–55, 62
  seismic testing, 14–15
  tire foundation, 39
Egypt, 6, 19, 22
electricity, 32, 120–21
emergency relief, earthbags used for,
  13–16, 19, 43
environmental building, xiv, 3–4
erosion, 31, 32, 75, 76, 77, 106
Escott, Carol, 16, 131–34
Europe, 9–10, 68. *See also* United
  Kingdom
expomat mesh, 25, 145, 146

extenders (fillers), 93, 95, 103, 105–109,
  114. *See also* sand
extensions, building, 17, 32–33
exterior finishes. *See* plasters
external features, 18, 30, 32

F
fiber. *See also* straw; straw-clay blend
  in earthen plasters, 77–78
  interior finish plaster, 92
  in lime plasters, 83–84
  in soil mix, 11, 106–109
fiber composite board, 109–10
fill. *See* earthbag fill
fillers. *See* extenders (fillers)
finishes. *See also* plasters; sealants;
      stabilizers
  casein, 98–100
  floor, 93, 117, 119–20
  interior, 92–100
  keying in, 76, 86, 87, 126, 133
  maintenance, 77, 91, 100–101, 109
  papercrete, 101
  roof, 66, 73
  soil mix for, 106–107
  spray application of, 76, 92
fire protection
  earthen plasters, 76
  insulation, 110
  thatch, 70
flood-prone areas, building in, 17, 30–
  31, 38, 43, 49
floors
  alternative, 118–19, 127
  earthen (*See* earthen floors)
Forschungslabor für Experimentelles
  Bauen, 15
foundations, 15–16, 34–41, 75, 76
  concrete, 127, 144–45
  in damp areas, 35–40
  details, 36–37
  in dry areas, 35–37
  dry-stone, 40
  earth-filled tire, 39
  examples, 127, 132, 136, 144–45

  flood-resistant, 30–31
  functions of, 34–35
  gabions, 30, 39–40
  ground, connection to, 35
  level plane provided by, 33, 47
  for non-earthbag buildings, 16, 33,
    38
  pumice-crete, 40–41
  rubble or mortared stone, 39
  site preparation for, 33–34
  trench, 35–38
  walls, attaching to, 38, 39
France, 9–10
furniture, 18, 30, 32, 103, 114–15
  finishes for, 75
  soil mix for, 106–107

G
gabions, 30, 39–40
geodesic structure, 130
glass, 139
glue, casein, 82, 97, 98–99
Gourmet Adobe, 82, 90, 96
gravel, 13, 30, 35–38, 46, 49, 50
Great Britain. *See* United Kingdom
Great Wall of China, 9
greenhouse, 63
Guatemala, 15
gypsum, 81, 82, 86, 92, 114

H
Hartworks, Inc. (Kelly and Rosana
    Hart), 12, 16, 130, 135–39
Hermosillo project, 16, 25, 56, 143–47
Hesperia Museum/Nature Center, 14,
    24, 34
hot climates, 31–32, 126, 132–33, 142
Howes, Dominic, 16, 125–29
Hunter, Kaki, 16, 52, 131, 140–41
hybrid construction
  Ahlquist square house, 127
  earth and straw bale, 112–14
  Escott-Kemble house, 16, 131–34
  foundations, 16, 33, 38–41
  greenhouse, 63

Hart dome house, 135–39
Howes conventional-style house, 128–29

I

insulation, 142
  earthen floor, 110, 111, 116
  foundation, 37, 38
  roof, 70, 72–73
  scoria as, 136
  walls, 50, 110, 112–14, 139, 145
interior finishes, 92–100
interior partitions, 103, 109–10, 114–15

J

jar test, 104–105

K

kaolin, 5–6
Kemble, Steve, 16, 131–34
Kennedy, Joseph F., 15, 16, 18
keying
  of earthbag walls with barbed wire/
    branches, 36, 54, 137
  finishes, keying in, 76, 77, 86, 87, 126,
    133
keystone, 23, 24, 59–60
Khalili, Nader, xiv, 14, 15, 21–22, 26, 34,
  38, 43, 124, 143
Kiffmeyer, Doni, 16, 52, 131, 140–41

L

lancet arch, 22–23, 44
landscaping, 30, 31, 32
lime-based stabilizers, 38, 46, 78–79,
  81–83, 86
lime from coral reefs, 142
Lime in Building: A Practical Guide
  (Schofield), 87
lime mortar recipe, 88
lime plasters, 66, 83–88, 136, 139
  alkalinity of, 83, 84
  application, 86, 87
  capping earthen plasters, 46, 77, 80,
    92, 93, 100, 114

casein paint on, 99
making, 86–87
permeability of, 82–83, 90, 93
pozzolanic additives to, 87–88
recipes for, 88
water resistance of, 75, 90
lime putty, 97
  gypsum added to, 92
  making (slaking), 84–86
  mixing with sand, 86, 87
limewash, 95, 97–98, 100
linseed oil
  lime-wash additive, 97–98
  in oil-based paints, 100
  as sealant, 93, 117, 119
  as stabilizer, 81, 89
lintels, 18, 62
living roofs, 65, 70–71, 140, 141
love, role in designing of, xvii

M

maintenance
  earthen floors, 119–20
  finishes, 77, 91, 100–101, 109
Malawi, 142
manure
  lime/manure render, 88
  manure/wheat flour/sand plaster, 82
Mexico
  Canelo project, 71–72, 100, 116
  Hermosillo project, 16, 25, 56, 143–
    47
  Obregon project, 25
  Save the Children project, 108–109
mica, 92
Middle East, 9, 19, 22, 25, 32
mineral stabilizers, 79, 82–83, 106
moisture
  barrier, 30
  damage caused by, 90–91
  earthbag fill, moisture content of, 46
  earthbags, moisture wicked into, 17,
    30
  permeability/evaporation of (*See*
    walls, breathability)

resistance to (*See* water resistance)
montmorillonite, 5–6

N

nailer boards, 114–15
natural building, xiii, 4
Njaya backpackers' lodge, Malawi, 142
Nubian vault, 26, 59

O

Obregon project, Mexico, 25
off-the-grid, 32, 126
oil, 81, 115, 117, 119. *See also* linseed oil
oil-based paints, 100
OK OK OK Productions, 16, 140
openings, 26, 54, 57–60, 138–39. *See*
    also compression rings
  arched, 13, 14, 21, 22, 58–60
  forms used for, 58–59
  in hot climates, 110, 126
  square, 57, 58, 60
Othona Community Retreat, xiv
Outram, Iliona, xiv, 14
ovens, 30, 75, 103, 114–15
overhangs, 32, 75, 76, 80

P

paints, 94–100
  casein, 99–100
  clay slip (*alis*), 95–97, 114, 141
  limewash, 95, 97–98, 100
  oil-based, 100
papercrete, 75, 135, 136, 138, 139
  plaster mix recipe, 12
  properties of, 11–12, 101
  as roof covering, 65, 66
parapet-tie wall buttresses, 23
passageways, 24–25, 136, 139
  connecting, 13
passive solar. *See* solar energy
pH, 81, 83, 84, 92, 106
plasters, 16, 19, 25, 126, 139. *See also*
    earthen plasters; lime plasters
  application of, 91–92
  for burlap bag construction, 76

plasters, *continued*
    fragmentation of render mass, 91
    insulative, 111, 136
    interior finishes, 92
    moisture permeability of, 76, 80, 90–
        91, 93
    for polypropylene construction, 75–
        76
    sand/manure/wheat flour, 82
    soil mix for, 106–107
plates. *See* bond beams
plumbing, 32, 120–21
polypropylene earthbags, 43–45
    biodegrading, 17
    cement stabilization of fill, 89
    finishes for, 75–76
porches, 80
Portland cement, 78, 81, 89, 139
potassium silicate, 94
pozzolanic material, 87–88. *See also*
        pumice
prickly pear juice, 81–82
Pueblo De Sarmiento, 16, 25, 56, 143–
    47
Pueblo Indians, 4, 5, 7, 90, 108
pumice, 37, 38, 49, 50, 73, 110, 114
pumice-crete, 38, 40–41

Q
quark, 88, 99

R
rammed earth, 5, 6, 9–10, 38, 76
rammed straw, 111
reinforcement rods, 25–26, 38, 54, 142,
        145, 146
remote locations, earthbag use in, 13,
        43
renders. *See* plasters
roofs, 65–73, 80
    adobe/brick, 66
    conventional, 68, 127
    earthbag foundation for, 33
    insulation, 70, 72–73
    living, 70–71, 140, 141

low-cost flat, 71–73
metal, 145
surface finishes, 66
thatched, 68–70
timber poles, use of, 135, 138, 139
vaulted, 22, 66, 67
water-catchment, 68
wind-resistant, 133

S
sand, 19
    in earthbags, 13, 35, 46, 49, 89, 136–
        37, 140, 142
    in plasters, 82, 83, 86–87, 92, 139
    in sealant, 93
    in soil mix, 103, 105–109, 114
Save the Children Foundation project,
        108–109
scoria, 38, 50, 130, 135, 136, 137–38
screed boards, 118
screens, use of, 32, 80
sculpting. *See* furniture
sealants, 80, 92–94, 115, 117, 119–20
seismic activity. *See* earthquake
        resistance
Serious Straw Bale (Bergeron and
        Lacinski), 70
sick building syndrome, 79, 95
site
    landscaping, 30, 31, 32
    locating building on, 13, 29–32
    preparation, 33–34, 47
    topography, 30–31, 34
slaking, 84–86
sloping site, 34
Smith, Michael G., 70, 104
sodium silicate, 93–94
soil for building materials, 80, 103–9.
        See also clay-based building
        materials; earthbag fill; sand
solar energy, 31, 110, 127, 128, 139
solid buttresses, 23
South Africa, 16, 18
southwestern United States, 6–7, 11–12,
        14, 19, 90, 123–27, 130, 135–39

spring line, 23, 26
square construction, 127
square openings, 57, 58, 60
squinches, 27
stabilizers, 46, 78–83, 98, 106. *See also*
        cement, finish stabilizer;
        cement-stabilized earthbags;
        sealants
    application of, 91–92
    lime-based, 38, 46, 78–79, 81–83, 86
    for waterproofing, 88–89
Steen, Athena and Bill, 11, 108–109,
        116, 119
stem walls. *See* foundations
stoves, 30, 75, 103, 114
straw
    in earthen plasters, 77–78
    for insulation, 110
    rammed, 111
    soil mix using, 103, 105–109
straw bale construction, 4, 5
    earthbag foundation, 16, 33, 38
    finishes, 76, 80
    of furniture, 114
    hybrid earth and bale, 112–14
    for insulation, 72–73, 111–14
    roofs, 71–73
straw-clay blends, 6, 10–11, 37. *See also*
        cob
    for furniture, 114
    for insulation, 72, 110, 111
string lines, 34
structural testing, 14–15
stucco, 90, 123, 127
subfloor, 117
sun, location related to, 13, 31, 80
Superadobe, xiv, 16
sustainable building, xiv, 4
Sustainable Systems Support, 16, 131

T
tamping, 47, 52–54, 139, 140
Tanzania, 142
Tassencourt, Shirley, 123–26
tension rings, 26, 36–37, 54–57

tests, soil, 104–105
thatched roofs, 68–70
thermal mass, 31, 50, 91, 111, 113
Three-Vault House, 16, 25, 56, 143–47
tie bars, 24
timber, use of, 33, 62, 135, 138, 139, 144
tires, earth-filled, 39, 111
Tlholego Learning Centre, South Africa, 18
tools, 46–49
topography, 30–31, 34
trust in material, value of, xvii–xviii

U
United Kingdom, 8, 68, 76, 107–108
Utah, 140–41
utilities, 32, 120–21, 139

V
Vaughan, Sue, 130
vaults, 14, 22, 24–26
    earthquake resistance, 15
    forms used for, 25, 59, 144–46
    Nubian, 26, 59

openings to, 58–59
for roofs, 22, 66, 67
three-vault house, 16, 25, 56, 143–47
width to length, ratio of, 25, 67
vegetable stabilizers, 79, 81–82, 106
ventilation, 32, 91, 139, 143–44

W
walls. See also thermal mass
    breathability (permeability), 46, 76, 79, 80, 82–83, 90–91, 93, 95, 106
    fragmenting, preventing, 137
    interior partitions, 103, 109–10
    retaining, 30, 39
    straight, 55–56, 60, 127–29
water added to earthbag fill, 46, 47
water-catchment roofs, 68
water glass, 93–94
waterproofing, 32, 36–37, 60, 75–76, 80, 141. See also drainage; finishes; flood-prone areas, building in; overhangs; water resistance
bricks, 66

damp-proof membrane, use of, 38, 60, 70, 136
earthen floor, 117, 119–20
roofs, 66, 68–70, 73
sealants, 93
stabilization for, 88–89
water resistance. See also waterproofing
    clay-based soil mix, 106
    earthen floor, 119
    papercrete, 12, 75, 101, 136
    stabilizers for, 78–83
wattle and daub, 4, 6, 10
waxes, 115, 119–20
weatherproofing. See waterproofing; water resistance; wind activity
wheat flour paste, 81, 82, 96, 97
whitewash. See limewash
wind activity, 32, 54, 62, 80, 132–33
windows. See openings
Wisconsin, 128–29

Y
Yaqui house, Mexico, 16, 25, 56, 143–47

OTHER RECENT AND RELATED
REAL GOODS
SOLAR LIVING
BOOKS

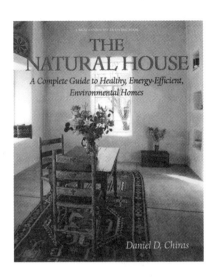

THE NATURAL HOUSE
A Complete Guide to Healthy, Energy-Efficient, Environmental Homes

Daniel D. Chiras

paper • 8 x 10 • 480 pages •
illustrations • b&w photos •
appendices • index
ISBN 1-890132-57-8 • **$35.00**

This exciting book, written by a veteran author who himself lives in a straw-bale and rammed-tire home, takes the reader on a tour of thirteen natural building methods, including straw bale, rammed earth, cordwood masonry, adobe, cob, Earthships, and more. You'll learn how these homes are built, how much they cost, and the pros and cons of each method. A resource guide offers contacts and resources for every region.

In a writing style that is clear, engaging, and fun to read, the author shows how we can gain energy independence and dramatically reduce our environmental impact through passive heating and cooling techniques, and renewable energy. Chiras also explains safe, economical ways of supplying clean drinking water and treating wastewater, and discusses affordable green building products.

While Chiras is a passionate advocate of natural building, he takes a careful look at the "romance" of these techniques and alerts readers to avoidable pitfalls, offering detailed practical advice that could save you tens of thousands of dollars, whether you're buying a natural home, building one yourself, renovating an existing structure, or hiring a contractor to build for you.

About the Author

Dan Chiras holds a Ph.D. in biology and teaches courses on sustainability at the University of Denver and the University of Colorado. He has published five college and high school textbooks as well as books for general audiences. Chiras is an avid musician, organic gardener, river runner, and bicyclist who lives with his two sons in a passive solar/solar-electric home in Evergreen, Colorado.

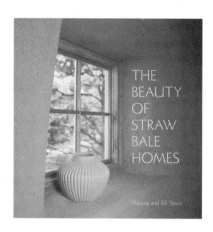

paper • 8 x 8 • 128 pages • color
ISBN 1-890132-77-2 • $22.95

In the past two decades, the bale-building renaissance has attracted some of our most gifted architects, artisan builders, and craftspeople. The hands-on process of constructing walls with a completely natural substance has appealed to both very experienced builders and those who find this to be a uniquely accessible form of creating shelter. The characteristic thick walls and wide windowsills of straw bale houses, the possibility of incorporating curves and even arches, and the rousing experience of family "wall-raisings" have become well-known. Combined with older styles of plastering and earthen floors, these very contemporary buildings have a timeless quality that's easy to recognize yet hard to achieve with conventional manufactured materials.

Athena and Bill Steen, co-authors of the original *Straw Bale House*, have now created a book that celebrates in gorgeous color photographs the tactile, sensuous beauty of straw bale dwellings. Their selection of photos also demonstrates how far bale building has come in a very short period of time. Along with handsome homes, small and not-so-small, this book shows larger-scale institutional buildings, including schools, office buildings, the Real Goods Solar Living Center, and a Save the Children center in Mexico.

In addition, this book includes an introductory essay by the Steens noting key lessons they have learned in years of building with bales: insights into the design and construction process, and critical advice about features that ameliorate the impacts of moisture, weather, and wear-and-tear over time. Each photograph is also accompanied by narrative text highlighting a given building's special features and personal touches.

About the Authors

Athena Swentzell Steen grew up in Santa Fe and at the Santa Clara Pueblo, building with natural materials from an early age. Bill Steen is a photographer and collaborative builder especially interested in combining building techniques with community-enhancing approaches to design. Athena and Bill live in Canelo, Arizona.

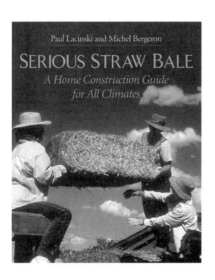

Paul Lacinski and Michel Bergeron

SERIOUS STRAW BALE
*A Home Construction Guide
for All Climates*

paper • 8 x 10 • 384 pages •
8-page color section • b&w photos •
illustrations • index
ISBN 1-890132-64-0 • **$30.00**

Bergeron and Lacinski's new book is the first to look carefully at the specific design considerations critical to success with a straw bale building in more extreme climates—where seasonal changes in temperature, precipitation, and humidity create special stresses that builders must understand and address. The authors draw upon years of experience with natural materials and experimental techniques, and present a compelling rationale for building with straw, one of nature's most resilient, available, and affordable byproducts.

For skeptics and true believers, this book will prove to be the latest word.

- Thorough explanations of how moisture and temperature affect buildings in seasonal climates, with descriptions of the unique capacities of straw and other natural materials to provide warmth, quiet, and comfort year-round.
- Comprehensive comparison of the two main approaches to straw bale construction: "Nebraska-style," where bales bear the weight of the roof, and framed structures, where bales provide insulation.
- Detailed advice—including many well-considered cautions—for contractors, owner-builders, and designers, following each stage of a bale-building process.

This is a second-generation straw bale book, for those seeking serious information to meet serious challenges while adventuring in the most fun form of construction to come along in several centuries.

About the Authors

Paul Lacinski is a partner in Green Space Collaborative, an environmental consulting firm offering integrated project management and innovative design services. He lives in Ashfield, Massachusetts. Michel Bergeron is a founding member of Quebec's legendary ecological design-build firm, Archibio. He lives in Montreal.

THE HOUSE THAT JACK BUILT

TREEHOUSES

DAVID PEARSON

Circle Houses: Tipis, Yurts, and Benders

The House That Jack Built series continues its exciting exploration into innovative housing styles with a foray into the world of portable houses. *Circle Houses: Tipis, Yurts, and Benders* looks at mobile homes from three wildly different cultural perspectives, each of which has increasing relevance in our contemporary world. Lots of color, great stories, and just enough how-to information to get you in trouble! This is a book that is guaranteed to stimulate your imagination.

cloth • 8 x 8 • 96 pages •
color photos • index •
ISBN 1-890132-86-1 • **$16.95**

The first book in THE HOUSE THAT JACK BUILT series, *Treehouses* introduces homemade buildings from all over the world, from tipis to treehouses to houseboats. Edited by David Pearson, best-selling author of *The Natural House Book* and *The Natural House Catalog*, these delightful books of practical inspiration will appeal to both dreamers and doers.

Treehouses are the epitome of fantasy. In this book, you will get enough practical information to get started, and enough inspiration to last a lifetime. Drawing from real projects from around the globe, the book features twenty distinctive tree dwellings. Each is a masterwork of inventiveness, due in part to the uniqueness of the respective settings, but also to the outpouring of creativity that accompanies the challenge of living in trees.

Beautiful color photographs portray finished projects and the construction process. Pragmatic tips and tricks provide less experienced builders with the confidence to begin designing their own treetop retreats. A how-to section illustrated with line drawings and easy-to-follow instructions gives even weekend carpenters enough information to complete a simple project. The extensive resource section offers places to visit, both real and virtual.

Thanks to Julia Butterfly Hill, who lived in a redwood tree for two years to protest logging practices in the California, treehouses have lately been prominent in the news. Their appeal is timeless. To live in the trees is to reside in a state of natural splendor. David Pearson, one of the pioneers in the natural building movement, captures the combined spirits of innocence and daring in this awesome little book.

Maybe the "house that Jack built" will become the house that you build!

About the Author

London-based architect David Pearson is author of *The Natural House Book*, *The Natural House Catalog*, and *Earth to Spirit: In Search of Natural Architecture*.

paper • 8 x 8 • 96 pages • color photos • b&w illustrations • index
ISBN 1-890132-85-3 • **$16.95**

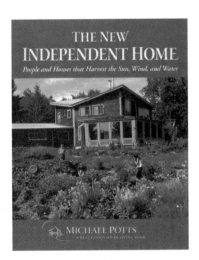

THE NEW INDEPENDENT HOME

People and Houses that Harvest the Sun, Wind, and Water

MICHAEL POTTS
A REAL GOODS SOLAR LIVING BOOK

paper • 8 x 10 • 400 pages •
illustrations • b&w and
color photos • index •
ISBN 1-890132-14-4 • **$30.00**

Michael Potts's 1993 book, *The Independent Home,* has become one of our best sellers and is flagship of the series of popular Solar Living Books, produced in association with Real Goods. Because of its impact in bringing the almost unknown promise of solar energy to thousands of readers, one longtime observer of energy trends described the publication of the original *Independent Home* as "the most important event in the solar industry in more than a decade."

In this newly revised and expanded edition, Potts again profiles the solar homesteaders whose experiments and innovations have opened the possibility of solar living for the rest of us. Potts provides clear and highly entertaining explanations of how various renewable energy systems work, and shows why they now make more sense than ever. He is a brilliant guide to the stages of planning and design faced by everyone who seeks to create a home that reconciles the personal and global dimensions of ecology.

Over the past five years, the concept of an "independent" home has evolved beyond the energy system to encompass the whole process of design and construction involved in planning a renovation or new home. Independent homes are homes with integrity and personality. Traditional-style dwellings are now being built with natural materials, such as straw bales and rammed earth, and combined with state-of-the-art electronic technologies for harvesting free energy from the surrounding environment.

Potts writes lucidly about homeowners some would consider to be the lunatic fringe, and others would regard as among the only sane people on the planet. This movement for self-reliance is an important trend that will continue to surprise and delight us as we approach the inevitable dawn of the Solar Age.

About the Author

Michael Potts is a freelance writer, networker, computer applications designer, and director of the Caspar Institute in northern California. He and his family have been building their own innovative, low-impact, off-the-grid home for twenty years.

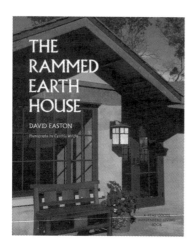

THE
RAMMED
EARTH
HOUSE

DAVID EASTON

Photographs by Cynthia Wright

A SELF-RELIANCE
INDEPENDENT LIVING
BOOK

paper • 8 x 10 • 288 pages •
color photos • b&w photos •
illustrations • resources • index •
ISBN 0-930031-79-2 • **$30.00**

Humans have been using earth as a primary building material for more than ten thousand years. As practiced today, rammed earth involves tamping a mixture of earth, water, and a little cement into wooden forms to create thick, sturdy masonry walls. Earth-built homes offer their inhabitants a powerful sense of security and well-being, and have a permanence and solidity altogether lacking in so many of today's modular, pre-fabricated houses.

About the Author

David Easton has led the revival of rammed earth in this hemisphere. He makes his living as a builder in northern California.

The Rammed Earth Renaissance

video • 45 minutes • ISBN 0-9652335-0-2 • **$29.95**

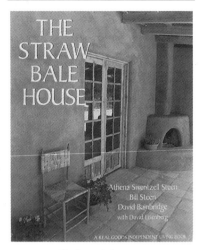

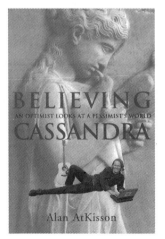